Vincent D Harris

The Diseases of the Chest Including the Principal Affections of the Pleurae Lungs, Pericardium, Heart, and Aorta

Vincent D Harris

The Diseases of the Chest Including the Principal Affections of the Pleurae Lungs, Pericardium, Heart, and Aorta

ISBN/EAN: 9783744693929

Printed in Europe, USA, Canada, Australia, Japan

Cover: Foto ©berggeist007 / pixelio.de

More available books at **www.hansebooks.com**

THE
DISEASES OF THE CHEST

INCLUDING THE

PRINCIPAL AFFECTIONS OF THE PLEURÆ, LUNGS, PERICARDIUM, HEART, AND AORTA

BY

VINCENT D. HARRIS, M.D.Lond., F.R.C.P.,
PHYSICIAN TO THE VICTORIA PARK HOSPITAL FOR DISEASES OF THE CHEST, ETC. ETC.

WITH FIFTY-FIVE ILLUSTRATIONS

PHILADELPHIA
P. BLAKISTON, SON & CO.
1012, WALNUT STREET
1888

PREFACE

In the present volume an attempt has been made to set out in a systematic and concise form the principal diseases of the chest.

The book is primarily intended for the use of students, but the Author hopes that it may be not without value to practitioners and others who may desire to recall to memory the chief known facts about the various diseases of the thoracic viscera.

In order to carry out the primary intention of the work, and to keep it from becoming too large for its purpose, every effort has been made to put the facts shortly and in tabular form, as well as to omit controversial matter and descriptions of diseases which are, although interesting and instructive to specialists, chiefly to be looked upon as clinical curiosities. For the same reasons the names of, and quotations from, authorities have been for the most part left out. The Author, however, takes this opportunity of acknowledging his indebtedness to the standard writers on chest diseases, many of whose books have been freely consulted.

Objection may possibly be taken to the amount of space devoted to the description of the normal chest.

From considerable experience of students, the Author has found that however well they may have mastered anatomy and physiology in the schools, the majority of them appear to be unable to apply their knowledge, in the physical examination of the chests of patients in the wards and out-patients' rooms. For this reason the insertion of a *résumé* of the chief points in the construction and working of the chest and its viscera likely to be of practical importance to them has been considered necessary.

The Author desires very heartily to thank Dr Andrew, for his encouragement and help as well as for the use of several tables drawn up by him; Dr Brunton, for permission to copy several tables from his 'Pharmacology;' Dr Tooth, for revising the chapter on the physical examination of the chest; and Dr Shore, who has not only devoted much time and trouble to revising and correcting the manuscript as well as the proofs of the book, but has also rendered invaluable help by his suggestions and criticisms.

Mr Danielsson has drawn upon wood and engraved with his usual skill, all of the original illustrations, except two or three copied by the Meisenbach Company. The original microscopic drawings were done by the Author from his own preparations.

 31, WIMPOLE STREET,
  CAVENDISH SQUARE;
   June, 1888.

CONTENTS

CHAPTER I

PAGE

THE NORMAL CHEST.—Plan of Construction—Of the Chest-Walls—The Dimensions and Contents of the Chest,—Trachea, Lungs, and Pleuræ—The Movements of Respiration—Action of the Lungs in Health—The Respiratory Centre and its Functions and Action—The Heart and Pericardium—Position of the Lungs and Heart with respect to the Chest-Walls—Innervation and Action of the Heart—Blood Pressure—Other Thoracic Viscera . . 1—46

CHAPTER II

THE SYMPTOMS OF THE DISEASES OF THE CHEST.—Hyperpnœa, Dyspnœa, and Asphyxia—Cough and Sputum—Hæmoptysis—Palpitation of the Heart—Angina Pectoris—Syncope. . . . 47—79

CHAPTER III

THE PHYSICAL EXAMINATION OF THE CHEST.—By Inspection, by Palpation, by Percussion, and by Auscultation—Stethoscopes—Stethometry, Spirometry, Pneumatometry, Cardiography, Sphygmography . . . . 80—121

CHAPTER IV

THE PULSE AND ITS INDICATIONS . . . 122—129

CHAPTER V

DISEASES OF THE PLEURA.—Inflammation or Pleurisy—Empyema—Paracentesis Thoracis—Hydrothorax—Diaphragmatic Pleurisy—Pneumothorax—Hæmothorax—Carcinoma and Tubercle of the Pleura
130—173

CHAPTER VI

DISEASES OF THE RESPIRATORY MUCOUS MEMBRANE.—Catarrhal Tracheitis and Bronchitis—Croupous Bronchitis—Bronchial Asthma—Bronchial Dilatation—Whooping-cough—Influenza—Hay Asthma
174—199

CHAPTER VII

DISEASES OF THE LUNG TISSUE PROPER.—Emphysema of the Lungs—Pneumonia, Croupous, Catarrhal, and Interstitial—Gangrene of the Lungs—New Growths of the Lungs and Pleuræ . . 200—241

CHAPTER VIII

PHTHISIS PULMONALIS.—Syphilitic Disease of the Lungs . . . . 242—293

CHAPTER IX

DISEASES OF THE PERICARDIUM.—Pericarditis—Adherent Pericardium—Hydro-pericardium—Pneumopericardium—New Growths of the Pericardium
294—308

CHAPTER X

DISEASES OF THE HEART.—Hypertrophy—Dilatation—Atrophy—Endocarditis—Myocarditis—Fatty Degeneration—Other Degenerations—Parasites—Rupture . . . . 309—341

CHAPTER XI

VALVULAR DISEASES OF THE HEART.—Aortic Disease—Mitral Disease—Tricuspid Disease—Pulmonary Disease—General Diagnosis and Treatment of Valvular Diseases . . . . . 342—374

CHAPTER XII

CONGENITAL DISEASES OF THE HEART . . . 375—377

CHAPTER XIII

DISEASES OF THE THORACIC VESSELS.—Aortitis—Atheroma—Aneurysm—Fatty Degeneration—Rupture of the Aorta—Constriction of the Aorta—Dilatation of the Aorta—Disease of the Pulmonary Artery—Pulmonary Embolism—Pulmonary Apoplexy—Disease of the Intra-thoracic Great Veins
378—400

CHAPTER XIV

INTRATHORACIC TUMOURS . . . . . 401—404

LIST OF ILLUSTRATIONS

	PAGE
1. Articulation of the sternum with the clavicle and ribs	3
2. Transverse section through the chest of the fœtus on level with the upper articulation of the eighth dorsal vertebra	10
3. Section of human bronchus (D. J. Hamilton)	13
4. Diagram of the position of the lungs and heart with relation to the chest wall. Front view	14
5. Ditto ditto. Back view	15
6. The heart within the opened pericardium lying upon the diaphragm	26
7. The heart as seen from the front	28
8. The heart as seen from behind	29
9. Diagram of the chest with the lungs *in situ*. Front view (Bourgery)	32
10. Diagram of the lungs within the pleuræ. From behind (Bourgery)	33
11. Diagram of the nervous mechanism of coughing (Brunton)	55
12. Plessor and pleximeter	91
13. Examples of wooden stethoscopes	97
14. Binaural stethoscope	98
15 and 16. Differential binaural stethoscope	98
17. Recording stethometer	115
18. Spirometer	116
19. Spirometer (modified from Hutchinson's)	117
20. Sanderson's cardiograph	118
21. Pond's sphygmograph	119
22. Mahomed's sphygmograph	120
23. Marey's tambour	121
24. Diagram of the extent of percussion dulness in right pleural effusion	142
24A. Diagram of the dulness in a case of right pleural effusion from behind	144
25. Posterior part of several intercostal spaces to show the relations of the intercostal arteries within them	155
26. Potain's trochar and cannula	156

LIST OF ILLUSTRATIONS

	PAGE
27. Trochar and cannula	156
28. Aspirator	157
29. Hensley's apparatus for tapping and washing out the chest	158
30. Diagram of the hyper-resonance in a case of pneumothorax	172
31. Bronchiectasis of the lung	187
32. Pulmonary emphysema (Ziegler)	204
33. Diplococci of pneumonia	215
34. Alveolus in the second stage of croupous pneumonia	217
35. Temperature chart of a case of acute pneumonia	221
36. Catarrhal or broncho-pneumonia	231
37. Segment of a rounded mass of consolidation in phthisis	258
38. A large miliary tubercle formed of the coalescence of two or more tubercles	259
39. A miliary tubercle	260
40. A giant cell	261
41. A nodule of softening interalveolar fibrous tissue with tubercle bacilli in the softened part	263
42. Section about half through a thickened pleura	264
43. Diagram of primary tubercle granulation of the lungs (Rindfleisch)	269
44. Tubercular sputum with bacilli	271
45. A heart and the sac of the pericardium, the latter opened to expose the inflamed surfaces	297
46A. Pulse tracing of pulsus paradoxus	300
46. Diagram of the circulation	310
47. Diagram of the enlargement of the heart in excentric hypertrophy	312
48. Section of a nodule in malignant endocarditis	326
49. A heart from a case of ulcerative endocarditis	344
50. Diagrams showing the effect of aortic valve disease upon the left ventricle	346
51. Pulse tracing in a case of aortic incompetence	350
52. Pulse tracing in a case of aortic incompetence	350
53. Diagrams showing the effect of mitral valve disease upon the left ventricle	353
54. Pulse tracing from a case of mitral incompetence	355
55. Pulse tracing in a case of aneurysm of the aorta	388

DISEASES OF THE CHEST

CHAPTER I

THE NORMAL CHEST

It is highly necessary before proceeding to the study of the **Diseases of the Chest**, that special attention should be directed to the **Normal position, structure, and functions of the Lungs, Heart, and** other thoracic viscera. It will not be out of place, therefore, if we devote the present chapter to recalling to the memory the most salient points in the anatomy and physiology of these organs. First of all, however, we shall find it useful to consider somewhat at length the way in which the chest walls are constructed.

The **Chest** or **Thorax** is a closed cavity which only indirectly communicates with the external air by means of the trachea. The cavity may be considered to have more or less the shape of a dome, somewhat flattened posteriorly. The framework of the walls of the cavity is formed by the ribs laterally, articulating with the dorsal spinal column behind, and by means of the firm but elastic cartilages, with the sternum in front. The posterior articulations of the ribs to the vertebræ are arranged in the following manner, in order that there shall be free movement. The head of each rib, with the exception of the first, eleventh, and twelfth ribs, articulates with the bodies of two contiguous vertebræ by two surfaces, the upper one being received into a facet on the lower part of the body of the vertebra above, the lower one into the facet on the upper part of the body of the vertebra below, whilst a slight ridge between the two surfaces is firmly fixed by means of an inter-articular ligament to the intervertebral disc. The two articulations of the head of the rib with the vertebræ are provided with ligaments, and possess complete synovial membranes, so that they must be looked upon as distinct joints. The heads of the first, eleventh, and twelfth ribs articulate only with their corresponding vertebræ, and each has one joint only. The tubercle of each rib, with the exception of the eleventh and twelfth, articulates with the transverse process of its corresponding vertebra by means of a joint provided with a synovial membrane and strong ligaments. The eleventh and twelfth ribs have no costo-transverse articulations.

The anterior attachments of the ribs may be thus briefly described (Fig. 1): the upper seven are attached by means of their cartilages to the sternum, the cartilage of the eighth joins the cartilage of the seventh, the ninth that of the eighth, and the tenth that of the ninth, thus sloping away from the middle line. The eleventh and twelfth ribs are free or floating, and are not connected to the others.

The costal cartilages of the upper seven or true ribs are received into corresponding depressions on the margin of the sternum, each forming a double joint provided with two synovial membranes, with the exception of the seventh cartilage, which has only a single joint.

The cartilages increase in length from the first to the seventh, and then diminish, that of the first being very short, and frequently ossified in the adult.

At the points of junction of all the cartilages of the ribs from the sixth to the tenth there are, generally speaking, synovial articulations.

The directions of the costal cartilages, as shown in the annexed diagram, should be noted. The first

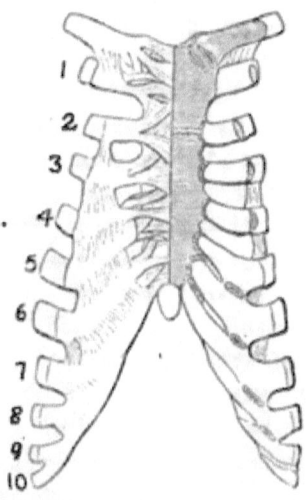

Fig. 1.—Articulations of the sternum with the clavicle and ribs, from the front. On the right of the middle line the anterior ligaments are shown. On the left side the front part of the sternum, clavicle, and costal cartilages has been removed, to show the articulations and synovial membranes; the rib-cartilages are numbered. (Modified from Allen-Thomson.)

descends slightly, the second is all but horizontal, and from the third to the tenth they all ascend, the third slightly, the fourth more so, and the others still more. It should also be noted that the **sternum** itself is made up of three bones, viz. the uppermost and broadest part or **manubrium**, the **body** or middle portion, and the lowest, the **xiphoid** or **ensiform cartilage**, and that these are not joined together by bone but by cartilage until late in life. The body is the longest portion, and consists of several segments, united at first by cartilage and later by bone. This condition of the sternum admits of some movement, which would hardly be the case with a rigid bone. From this cause also some of the deformities of the chest arise in cobblers, compositors, and the like. The upper two rib-cartilages articulate with the manubrium of the sternum, and the second to the seventh with the body of the bone, whilst the second fits into the cartilaginous junction between them, and articulates with both. At this articulation of the two portions of the sternum is a slight angular projection forward, much exaggerated in some individuals, forming in the covered chest the **angle of Ludwig**. By the arrangement of the articulations of the ribs which has been now described, **movements in two directions** are permitted, viz. those of **elevation or depression, and of rotation**; the fixed point in both cases being the attachment by the inter-articular ligament to the intervertebral disc. The first of these movements is simple, the rib at its costo-transverse joint moving upwards and backwards or forwards and downwards. The rotatory movement takes place around an axis joining the vertebral and sternal ends of the bone.

It should be remembered that the costal cartilages frequently ossify and become rigid in old age, and that the ensiform cartilage is ossified in the adult.

The following **landmarks** are also useful:—The top of the sternum is on a plane with the lower border of the second dorsal vertebra behind, the angle of Ludwig with the lower border of the fourth, and the junction of the body of the sternum to the ensiform cartilage with the lower border of the eighth, dorsal vertebra.

Openings of the Chest.—In the skeleton, the upper opening is narrow. It is bounded by the first dorsal vertebra behind, the upper bone of the sternum in front, and the short first rib on either side. The **lower opening** is broad; it extends from the ensiform cartilage to the twelfth dorsal vertebra. The shape of the chest in health may be inferred from the bony framework. Its antero-posterior and transverse diameters increase from above **downwards**, and its shape on transverse section is made irregular from the projection **forwards of the** spinal column (Fig. 2), which **renders its** antero-posterior **diameter** smaller than it would otherwise be, and smaller too than its **transverse diameter.** The shape of a transverse section of the chest at the level of each dorsal vertebra down to the seventh is reniform; on some levels as much as a fourth or a fifth of the whole area being behind a line drawn across the front of the body of the vertebra.

It is necessary now to say a few words about the intercostal muscles. The outer of the two layers, or **external intercostals,** are directed obliquely downwards and forwards, being attached to the outer edges of the ribs. They **extend from** the transverse processes behind to the end of the rib anteriorly, but are not found between the cartilages. They elevate and rotate the ribs. The inner layer, or **internal intercostals, begin** at the angles of the ribs, being directed from above downwards and backwards from their attachments at the inner borders of the ribs, and they extend forwards to the sternum. It is generally thought at the present time, **that** the intercartilaginous fibres act much in the same way as external intercostals, whereas the other fibres depress the ribs. It will be noticed that both behind and in front there is only a single layer of muscle between the ribs, whilst laterally there are two layers.

The **intercostal arteries and nerves** in their course from behind forwards in an interspace, lie first of all upon the external intercostals (Fig. 25) and then between the two layers of muscle beneath the lower border of the rib above, but a branch of the artery runs along the upper border of the rib below.

The **width of the intercostal spaces** in the adult is **about** three quarters of an inch, but may be increased to one and a quarter inches by raising the arm above the head. The width is less behind and at the sides than in front, **and may be greatly** diminished in diseased conditions.

The floor **of the cavity** of the chest is formed by the musculo-tendinous **Diaphragm** or **midriff**. It forms a very arched partition between the chest and abdomen, particularly when in a relaxed (or non-contracted) condition, allowing the abdominal organs to encroach, as it were, upon what is usually looked upon as the chest, that is to say, the space enclosed by **the ribs**. The muscle arises from the ensiform cartilage, from the inner surfaces of the six lower costal cartilages, and from two tendinous arches over the posterior abdominal muscles (quadratus lumborum and psoas), as well as from the bodies of the first three lumbar vertebræ by tendinous slips. The muscular fibres converge to a so-called *central tendon*, the highest part of the muscle. The muscular sides are greatly arched but become less so when the fibres contract, the **ribs being** at the same time elevated. The attachment of **the** diaphragm behind **leaves a** very narrow space between it and the posterior chest wall at its lowest part. Into this narrow space, as will be seen later on, the posterior inferior edge **of the** lung is received. The **highest level** which the diaphragm reaches **during ordinary expiration**, is the fifth rib on the right side and the sixth rib on **the** left side; whereas **during forced expiration** it **may reach** a rib above on either side, and during inspiration it descends one or two inches. Its antero-posterior inclination would be represented by a line from the ensiform cartilage in front to the tenth rib behind.

The intercostal muscles and the diaphragm are the **most important muscles of inspiration**. In addition **to these** should be recollected the levatores costarum, each extending from the transverse processes of the vertebræ obliquely to the rib below, their direction being the same as the external intercostals, and the subcostals which are found internal to and in the same direction as the internal intercostals, chiefly in the neighbourhood of the angles of the ribs; they extend over two spaces. The triangularis sterni should also be mentioned; it is continuous as it were with the diaphragm.

Other muscles attached to the chest are important from two points of view, viz. that they may be used in extraordinary inspiratory efforts, and also that they form a more or less complete support to the intrinsic muscles of the chest. It will be useful to mention them in the form of a table.

Raise the ribs.
- A. **Acting from the shoulders, fixed by the rhomboidei** and other muscles.
  - i. The pectoralis minor—attached to the third, fourth, and fifth ribs, near their cartilages anteriorly;
  - ii. Serratus magnus—attached to the first eight or nine ribs external to the pectoralis minor.
- B. Acting from the head, which is fixed by the muscles of the neck and back,—the sterno-mastoid, scaleni and other sternal muscles, and the trapezius.
- C. Acting from the raised arm, the shoulders being fixed,—the pectoralis major and latissimus dorsi.

The serratus posticus superior is probably a muscle of ordinary inspiration.

As a practical point it should be noted that by standing upright and straightening the spinal column the upper intercostal spaces are dilated, and so inspiration is aided.

Ordinary expiration is seldom to any extent a muscular act. Those muscles which may act then, and do act, energetically during **forced expiration**, are the internal intercostal and abdominal muscles aided by part of the triangularis sterni and by the quadratus lumborum and serratus posticus inferior.

Covering the bony and muscular framework of

the chest wall is a varying amount of fat and subcutaneous tissue, as well as, in females, the mammary glands, but some of the bony prominences in connection with the chest are practically subcutaneous.

External dimensions of the chest.—In the adult the measurements given by different authorities vary. The following measurements are of strong, well-built adults:

	Men.	Women.
Circumference under the arms	34·3 in.	32 in.
„ at the ensiform cartilage	32 in.	30·4 in.

These numbers we should consider considerably above the average.

The **long diameter** of the chest is very variable, it may be 8, 10, or 11 inches, or more.

The transverse diameter of the **covered chest*** in the upper and lower parts measures:

In men 9·84 in. to 10·23 in., and in women 9·05 in. to 9·44 in.; whereas the **antero-posterior** diameter varies from 6·29 in. to 7·48 in.

Before quitting the subject of the construction of the chest walls it is as well to draw attention to a more or less unprotected spot in the fifth space anteriorly beneath the nipple, which is covered by the internal intercostals, the intercostal fascia, and the

---

* The diameters of the bony thorax in the natural position, according to Holden, are as follows:

A. Antero-posterior—
 (1) Upper opening at line of first rib . 2¼ inches.
 (2) At articulation of the manubrium with body of sternum . . . 4½ „
 (3) At junction of the body with the ensiform cartilage . . . 5⅜ „

B. Transverse—
 (1) At upper aperture . . . 4⅜ „
 (2) Between second ribs . . . 7 „
 (3) Between third ribs . . . 8⅛ „
  increasing up to
 (4) Between ninth ribs . . 10⅝ „
  and decreasing below.

weakest portion of the pectoralis major, and external oblique muscles, together with the common fascia and skin only. It has been properly held to be of great importance in connection with the pointing externally of collections of pus within the pleura. Another unprotected spot in the chest wall is stated by Mr. Norman Porritt, in his very useful Monograph on 'Intrathoracic Effusions,' to be in the eighth and ninth interspaces a little external to the inferior angle of the scapula.

The upper part or apex of the dome-like cavity of the thorax is closed in, it will be remembered, in the recent state by the various muscles and other structures which go to form the neck, and which are either attached to the upper ribs and sternum or pass into the cavity itself. Of these latter one is the trachea, which occupies the central position and, as it were, divides the opening into two parts.

Having now considered at some length the formation of the chest and of its walls we next turn to the contents of the cavity.

Contents of the Thorax.—On either side of the chest, and occupying the greater part of its area, is the **Lung**, covered by the serous sac of the pleura. Between the lungs is a space called the mediastinum, in the middle of which, and occupying its greatest part, is the heart enclosed within its fibro-serous bag, the pericardium. The part behind it is called the **posterior mediastinum**, and the part in front the **anterior mediastinum**, that above the pericardium is also sometimes called the **superior mediastinum**.

The **posterior mediastinum** is limited in front by the pericardium and the roots of the lungs, behind by the dorsal vertebræ from the fourth downwards, and laterally by the lungs and pleuræ; it contains the descending aorta, the vagi, azygos veins, thoracic duct, and lymphatic glands.

The **anterior mediastinum**, a small space, narrow above where the pleuræ come into contact and broader below. It contains several small anterior mediastinal lymphatic glands.

The **superior mediastinum,** the space above the pericardium; the lower limit of which is considered to be the plane passing through the lower part of the body of the fourth dorsal behind and the junction between the manubrium and body of the sternum in

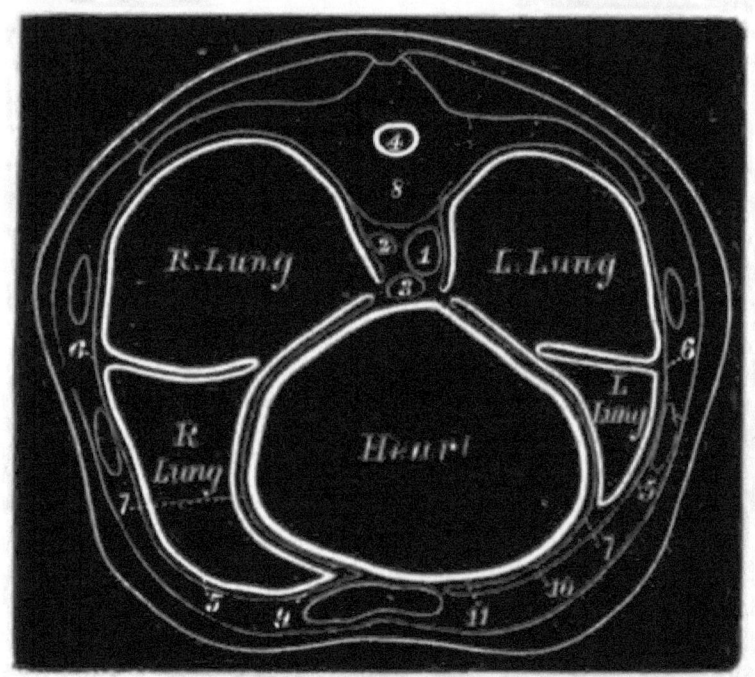

Fig. 2.—Transverse section through the chest of the fœtus, on a level with upper articulation of the eighth dorsal vertebra. 1, Descending aorta; 2, vena azygos; 3, œsophagus; 4, vertebral canal; 5, pleura pulmonalis; 6, pleura costalis; (7) pleura pericardii; 8, eighth dorsal vertebra; 9, sinus costo-pericardæus pleuræ dext.; 10, sinus costo-pericard. pleuræ sinistr.; 11, both layers of the pericardium. (Modified slightly from **Luschka.**)

front; it extends to the upper aperture of the thorax, and contains the trachea, œsophagus, and thoracic duct, the transverse portion of the arch of the aorta and its branches, the innominate vein and the superior cava, the phrenic, vagus, cardiac, and left recurrent laryngeal nerves, with lymphatic, and thymus glands.

Trachea, Lungs, and Pleuræ

The Trachea is a tube about three quarters of an inch in diameter and four and a half inches in length. Above it is continuous with the larynx, and below it divides into the two bronchi, at a level of the body of the fifth dorsal vertebra. The tube, it should be remembered, is supported by about eighteen cartilaginous rings, incomplete behind, which keep it open under variations of pressure. A fibrous membrane encloses the rings, and two layers of unstriped muscle, transverse and longitudinal, supply their place within the membrane behind. The mucous membrane is remarkable for the great amount of elastic tissue which it contains, and for a thick stratified epithelium, of which the upper layer is columnar and ciliated, laid down upon a basement membrane of considerable toughness. A certain amount of areolar tissue forms a scanty submucosa, separated from the mucosa by a longitudinal layer of elastic tissue. Within this coat are contained the mucous glands.

Of the two main bronchi, the right is somewhat shorter, wider, and more horizontal than the left. In structure the bronchi resemble the trachea. The cartilages of the main bronchi are imperfect rings behind as in the trachea, but when by dividing dichotomously the bronchi become smaller and intrapulmonary, the cartilages are irregularly disposed in plates around the tubes, and not in rings, so that no part is flattened as in the trachea and extra-pulmonary bronchi. The extra-pulmonary bronchi branch dichotomously into finer and finer subdivisions, which do not anastomose until the terminal bronchi or bronchioles are reached, and which open into alveoli or air passages beset with sacculations, and terminating in infundibula or blind terminal tubes with **air-vesicles**. The terminal bronchiole is about **one twenty-fifth of an inch in** diameter, cartilage is not found in their walls, neither are the **mucous** glands present beyond this point of the air tube. In structure the intra-pulmonary bronchi are akin to the **trachea.** They possess a mucous membrane, becoming thinner and thinner towards the terminal subdivision, and a complete muscular coat, traceable into the smallest **bronchi.** Of the mucous membrane in the medium-sized bronchi the epithelial cells, **as seen in the** accompanying figure, **are** arranged in **three layers.** The more superficial, columnar **and** ciliated, **the next** transitional, **and the** deepest flattened. These layers are placed upon a structureless, homogeneous, hyaline basement membrane, which Hamilton looks **upon as important in** imparting " the superficial character to the catarrhal affections of the bronchi." Next to the basement membrane comes the inner fibrous **layer** containing **many** elastic fibres, generally longitudinal, and **a** branching network of lymphatic vessels and spaces, after this the layer of unstriped muscle. Outside the muscle is another fibrous layer or adventitia, chiefly of fibrous tissue, but with **some** elastic fibres. This fibrous coat is continuous with the connective tissue surrounding the pulmonary arteries, and also, according **to** Hamilton, with the inter-**lobular** fibrous septa, which latter fact **he** considers important as explaining the formation **of** bronchiec-tases in old-standing cases of bronchitis. In **the** outer **coat** carbon particles often form a natural injecting **material** for the lymphatic spaces. Other particles **are continuous** with those spaces around the pulmonary artery, in the interlobular septa, then with the lymphatics **in** the deep layer of the pleura, and so with the bronchial glands at the base of the lung.

Tubes beyond one twenty-fifth of an inch in diameter chiefly consist of an external fibro-elastic layer and an internal muscular layer lined with short ciliated cells. These cells, whose cilia are said to move in an outward direction, give place to cubical epithelial cells when the minute tubes become sacculated to

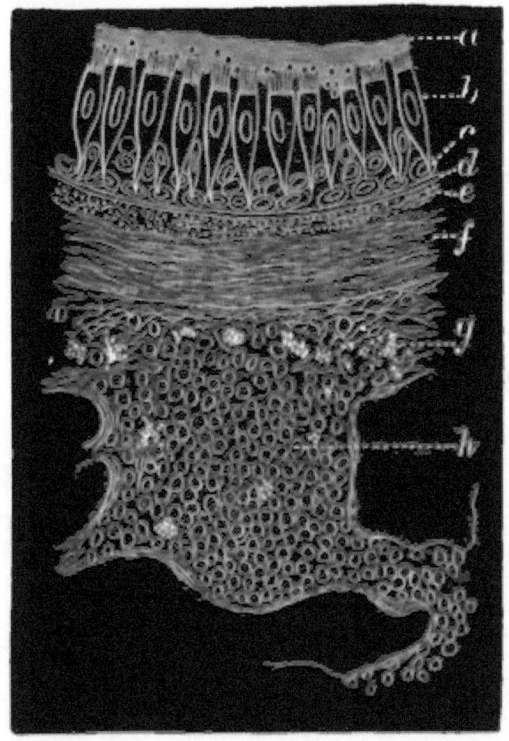

FIG. 3.—Section of human bronchus. *a*, Indicates the mucus precipitated upon surface; *b*, ciliated columnar epithelial cells; between *b* and *c* the transitional layer; *c*, deep germinal layer of cells (Debove's membrane); *d*, elastic basement membrane; *e*, inner fibrous coat; *f*, muscularis; *g*, outer fibrous coat, with *h*, lymph-adenoid tissue in it, and pigment granules. (After D. J. Hamilton.)

form air-vesicles. The presence of unstriped muscle throughout the whole of the bronchial system of tubes is to be specially noted, as well as the resisting basement membrane upon which the epithelium is laid down. The mucous membrane is also provided with a considerable amount of adenoid or lymphatic tissue, and with tubular mucous glands, as well as with copious blood- and peribronchial lymphatic-vessels.

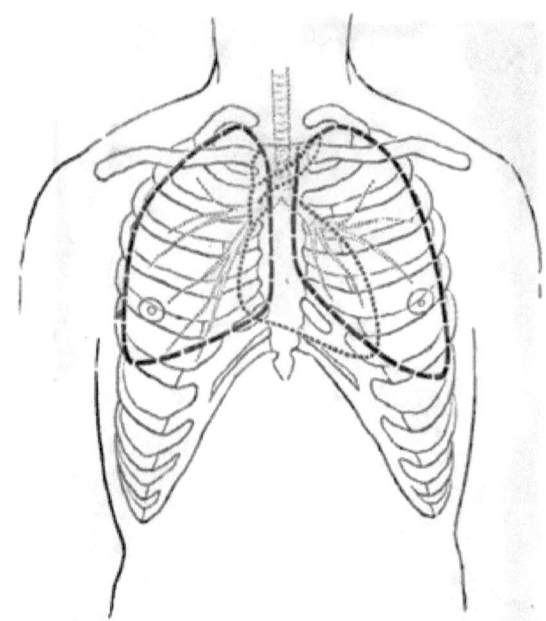

FIG. 4.—Diagram of the position of the lungs and heart with relation to the chest wall. **Front view.** The thick interrupted lines represent the boundary of the lungs, with their bronchi on either side in faint dotted lines. The heart and vessels are dotted in by thicker dots. The pleuræ extend more anteriorly to the lung-line, meeting behind the body of the sternum.

The **Lungs**, one on either side, are attached as it were to the main bronchi (Figs. 4 and 5). Although they differ slightly from one another in form, each may be said to be shaped somewhat like half of a bisected cone, with the cut surfaces facing one another. The base is excavated and rests upon the diaphragm, and its posterior border extends lower than the anterior, being received into the somewhat narrowed interval, to which attention has already been drawn, between the ribs and the arch of the diaphragm behind, and to the outer side. The outer surface is broad and convex, and extends from the posterior

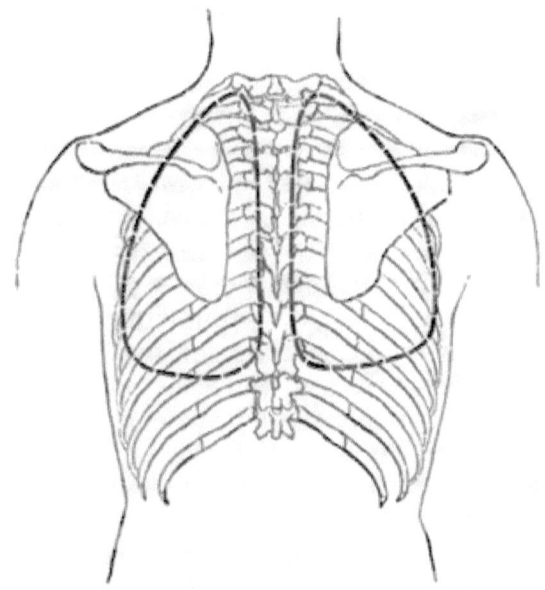

Fig. 5.—Diagram to represent the relation of the lungs to the chest wall. **Back view.** The interrupted black lines represent the limits of the lungs. The pleura must be supposed to extend further downwards.

to the anterior border. The inner surface is concave and partially covers the pericardium.

The apex of the lung is blunt, and it extends about an inch and a half above the first rib. The posterior border is thin, and is received into the interval between the ribs and the spinal column. The anterior border, much shorter than the posterior, is also thin, and in part overlaps the pericardium. The bronchi and vessels which form the root enter each lung at the inner side less than half way down, and somewhat more posteriorly than anteriorly.

Each lung is divided into lobes, the right into three and the left into two, by means of fissures. These fissures extend in the right lung as follows. The superior fissure runs from somewhat above the middle of the external surface of the lung to the root, the inferior from very nearly the bottom of the same surface to the root; between these two fissures is included the smallest or middle lobe. In the left lung the fissure passes upwards and inwards towards the root from the external surface, and separates the upper and smaller from the lower almost quadrilateral lobe.

The left lung is deeply notched anteriorly, where the heart and pericardium are situated, and so differs from the right in this respect as well as in having only two lobes; it is also somewhat narrower but longer than the right, because of the position of the heart on the left side and the liver on the right side, the diaphragm in consequence rises slightly higher on the right side than on the other.

Divisions of the bronchi, with respect to the lung, take place thus: The undivided portion of the right bronchus measures about an inch in length, gives off a branch to the upper lobe which soon divides into two; slightly lower down it gives off a branch to the middle lobe, whilst the remainder of the tube passes to the lower lobe, where it soon breaks up. The last branch is in the most direct line with the trachea, and is in fact its continuation. The left bronchus may be said to divide into two parts, one to the upper, and the other to the lower lobe, one being scarcely more a continuation of the trachea than the other.

Appearance in health.—When the chest is opened the lungs appear to occupy only a third or less of the space within. This is due to their collapse, which arises from their elasticity.* They vary in

---

* The elasticity of the lungs, as estimated by tying into the trachea a tube in connection with a water manometer and then opening the thorax, is equal to a column of about three inches

colour from a light reddish pink to a dark red, being grey in adult life, spotted or mottled from pigmentation. The tissue is soft and spongy, crepitating on cutting or on squeezing, and floating in water. The right lung weighs twenty-two ounces, whereas the left lung weighs only twenty ounces.

The Pleuræ form the serous investment of the lungs. Each pleura is quite independent of the other, and consists of a closed sac, the opposed surfaces of which are practically in contact. Each sac may be described as consisting of two layers, namely a parietal layer, which externally is adherent to the chest wall and forms its inner lining, and below is adherent to and lines the upper part of the corresponding half of the diaphragm, and a visceral layer which is reflected over and is adherent by its internal surface to the corresponding lung, a small amount of subpleural connective tissue only intervening. Above and below the root of either lung the parietal and visceral pleuræ are in contact, and the portion below passing downwards to the diaphragm as a double layer is sometimes called the *ligamentum latum pulmonis*.

Extent of the Pleura.—The limits of the pleura may be mapped out as follows:

a. **In the vertebral line**—

On the right side.	On the left side.
The lowest limit reaches the twelfth rib at its junction with the twelfth dorsal vertebra.	Do.

b. **In the mid-axillary line**—it reaches

The lower edge of ninth rib.	The lower edge of tenth rib.

---

of water. The elastic tension of the thoracic wall, as Douglas Powell points out, must be equal and opposite to this, and must have a constant tendency to expand the lungs, except when the thoracic parietes are fixed and inelastic. If this be true the stretched condition of the chest walls should be increased in expiration.

c. In front—it reaches

    The seventh costal cartilage at its junction with rib.   Rather below the junction of the seventh rib and its cartilage.

The pleural sacs are in contact in the mid-sternal line behind the second bone of the sternum; below this they diverge, the left shelving off suddenly, the right continuing in the same line for some distance. It is a matter of importance to remember that the extent of the pleura forwards and downwards is greater than that of its corresponding lung, except possibly after the deepest possible inspiration.

Pleural fluid.—In health there is a small quantity (2 to $3\frac{1}{2}$ fluid drachms) of a serous fluid within each pleural sac to lubricate the surface, so as to allow free movement of one surface upon the other. The fluid is alkaline in reaction, of sp. gr. 1012—1020, and contains one or more proteids (serum-albumin, and paraglobulin) besides fibrinogen, as well as chlorides, phosphates, and carbonate of sodium, and possibly a small quantity of glucose. It is not, as a rule, spontaneously coagulable.

The roots of the lungs consist of a bronchus, with branches of the pulmonary and bronchial vessels, nerves, lymphatics, and glands, &c., enclosed in areolar tissue and covered by the pleura. The superior vena cava lies in front of the root of the right lung, and the root of left lung passes in front of the descending aorta. The pulmonary plexuses of nerves lie one in front and the other, the larger, behind the root. The phrenic nerve is in front, and the vagus behind. The relation of the structures in each root is as follows:—From before backwards, branches of the pulmonary veins, next of the pulmonary artery, and posteriorly the bronchus with its vessels, nerves, and lymphatics. On the right side the bronchus is above, and on the left side below, the corresponding branch of the pulmonary artery, but the pulmonary veins are lowest in each case.

Minute structure of the lungs.—The whole lung is made up of lobules, which are the sacculated terminations of the ultimate bronchial tubes and infundibula, which have been previously mentioned. When the small tube opens into an infundibular passage, that is to say the part which is sacculated, the lining of fine cubical or small columnar ciliated cells is replaced by flattened epithelium, but here and there in the pounched-out air-vesicles there are found groups of small cells which represent these columnar cells. The walls of the air-sacs are formed of rudimentary areolar tissue supported by some elastic and muscle-fibres.

The blood-supply of the air-tubes and lungs.—The trachea is supplied with blood from the inferior thyroid artery, which forms a plexus of capillaries from which the blood is collected into the thyroid veins. The bronchi are supplied with blood by the bronchial arteries, except near the alveoli, where pulmonary branches give off some small offshoots for their supply. The meshwork of the capillaries in the mucous coat is fine; it is not so fine in the outside of the muscular coat. The distribution of the pulmonary artery, as it brings the venous blood to the alveoli to be aerated by contact with the alveolar air, is very important. The first branch to each lung follows in its divisions those of the corresponding bronchus, except that towards the alveoli the arterial branches divide more frequently. The main peculiarity about the manner of distribution of each main subdivision is that its branches do not anastomose either among themselves or with the neighbouring vessels. The terminal branch to an alveolus lies between it and its contiguous alveolus, and runs round its broadest part, and from it arises a narrow-meshed capillary system covering the whole alveolus, being separated from the interior by a single layer of flat cells only. From this capillary meshwork arise the radicles of the pulmonary veins. It should be remembered that the bronchial vessels supply the lung-substance, forming capillary plexuses not only on the bronchial tubes and in the coats of the arteries, but also in the interlobular connective tissue and upon the surface of the lung. Part of the blood from the bronchial arteries enters the pulmonary system; the remainder is collected into the right and left bronchial veins, which discharge themselves into the azygos veins. It is doubtful whether part of the smaller radicles of the bronchial vein do not enter the corresponding pulmonary veins. The bronchial arteries supply the bronchial glands.

It is interesting to note that the tracheal and bronchial arterial supply, although connected, are quite independent. It is also important to recollect that the bronchial circulation is really part of the systemic, and not of the pulmonary circulation.

The **nerve-supply** of the lungs and air passages is practically similar. It is derived from the vagus and sympathetic, and their branches; the **superior laryngeal** nerve being supplied to the **laryngeal mucous membrane**, the inferior **laryngeal**, the sympathetic and branches of the vagus trunk to the trachea, and the anterior and posterior **pulmonary plexus, made up of the branches of the vagi and sympathetic** nerves to the **bronchi** and lungs.

The **lymphatic system** of the lungs begins in the alveoli, in the walls of which are stomata opening into the alveolar lymph lacunæ. These lacunæ are the beginnings of the lymphatic vessels surrounding the pulmonary blood-vessels, the so-called **perivascular** lymphatics; they anastomose with the system of lymphatics around the bronchi, the **peribronchial**. The superficial lymphatics of the lung pass into a system of **subpleural** lymphatic vessels with valves. All of these lymph vessels ultimately run towards the bronchial glands. Direct communication thus exists between the interior of the alveoli, the lymphatic system, and the bronchial glands, and in this way is explained the implication of those glands when the alveoli contain morbid material. There is much lymphoid tissue, in the mucous membrane of the **larynx and** bronchi especially; smaller collections are **scattered** about the alveoli of the lung, near the smaller bronchi, **and underneath** the pleura. These collections of lymphatic tissue probably play an important rôle in the pathology of true tubercle.

The movements of respiration.—The respiratory action of the ribs has been already mentioned, but it is necessary now to return to the subject, for unless the natural action of the chest be understood in its entirety, abnormal performance of the various factors concerned cannot be gauged even if noticed. Inspiration, **or the taking** in of air, let it be remembered, is a muscular act, generally automatic, but capable of being influenced reflexly, as well as by the will—is usually performed by the diaphragm, assisted by the lower intercostal **muscles.** This is the case with men. With women it is somewhat different; their upper intercostal muscles always **take** a prominent part in the process. Thus **we have two** types of respiration, **the former costo-abdominal** or **inferior costal, and the latter costal** or **superior costal.** In infants, the diaphragm acting almost alone, constitutes the **abdominal** type. As we have before indicated, an **increase** in the area of the chest when all the muscles act takes place in all the diameters of the cavity, viz. in the antero-posterior, by the elevation and straightening of the sternum; in the transverse by the elevation and rotation of the ribs; in the lateral diameter by the opening out of the angle of the rib and its cartilage, and in the vertical by the descent of the diaphragm.

Expiration, which follows the inspiration without any interval, is, **as a rule, not a muscular act, but is**

due to the elastic recoil of the chest wall and lungs. When expiration is obstructed, the muscles of expiration and of extraordinary expiration are called into action, and when acting strongly are found to be even more powerful than those of inspiration.

Following expiration is a slight pause. Expiration is longer than inspiration as 8 to 6. There should be 16 to 18 complete respirations a minute, but the number varies a little at different ages.

The lung in a healthy condition follows every movement of the chest walls. The upper part appears to move the least. The expansion of the lung takes place in two directions, from above downwards and from behind forwards; the fixed points in the two movements are at the apex and a point on the posterior surface of the lung respectively. If the pleural cavity be perforated from within or from without the lung collapses. If a part of the lung refuses to expand and aid respiration, there may be corresponding depression of the chest wall.

Action of the lungs in health.—The object of respiration may be said to be twofold, viz to oxygenate the venous blood and to eliminate from it various effete substances, of which the chief and most important is carbonic acid gas. It is hardly necessary to do more than mention that these two objects are attained by gaseous diffusion, the air within the lungs being renewed by the alternate enlargement and contraction of the chest in inspiration and expiration. As the lungs follow the movements of the chest wall and expand, they draw in fresh air through the trachea, and again by contraction expel a certain amount of already vitiated air, the difference between the oxygen percentage of the expelled from that of the inspired air in average ordinary respiration being nearly 5, and of the carbonic percentage of the expired over the inspired air being about $4\frac{1}{2}$.

Now, the inspirations of each individual under ordinary circumstances in health are practically constant as regards number and depth, and we must therefore suppose that in health the total quantity of oxygen required for the due performance of all the bodily functions in a given time is obtained by a certain number of inspirations of a given depth— and the same applies to the carbonic acid got rid of by expiration. So that whatever respirations lose in depth they gain in number, and *vice versâ*.

The regulation in the number and depth of the inspirations depends upon the influence of the nervous system, and especially upon the regular as well as sufficient stimulation of the respiratory centre in the medulla, the so-called nœud vital of physiologists, which corresponds in position with the vagus nuclei.

The stimulus which excites the respiratory centre proper and the subsidiary centres in the spinal cord, if they exist, to action, is the blood which circulates through them. This statement accepts without discussion the belief in the automatic action of the respiratory centres, for it appears to be based upon sufficient experimental evidence. But if the centres be automatic there is also abundant proof that the medullary centre at all events is capable of being influenced by afferent impulses conducted to it by the vagus nerves, especially by those fibres which are distributed to the lung tissue, and the influence is shared to some extent by afferent impulses conveyed by other nerves.

The terminal filaments of the vagi in the lungs are stimulated either by the blood in the pulmonary capillaries or by the gases of the pulmonary alveoli, or by both.

It has been demonstrated by experiment that the quality of the blood which stimulates the medullary centre to action is its venosity, and also that it is the reduction of the oxygen of the blood below a certain standard which is effectual, and not the presence in it of an excess of carbonic acid gas. As long as the oxygen is maintained at a certain standard the centre is not stimulated, and no muscular effort of inspiration occurs, and so if the inspirations are deeper than normal they become less frequent, simply because more oxygen having been inhaled and taken in by the blood, a longer time elapses before enough of the oxygen is used up by the tissues to produce a reduction of its amount in the blood circulating in the medulla to what may be called the stimulation point. But as soon as this stimulation point is reached, the medullary centre not only sends out impulses of its own accord as it were, but it also energetically responds to afferent impulses.

A prolonged interval between two inspirations caused by an excess of oxygen in the blood is known as apnœa. The condition should not be confounded with **asphyxia**, to which the term is sometimes applied by medical jurists.

The respiratory centre, however, is apparently not inspiratory only in function. It is also expiratory in part, and from it are sent out, under certain conditions, efferent impulses which stimulate the expiratory muscles to contraction. In consideration of the respiratory movements, both in health and disease, this double action of the centres must not be disregarded. The reflex, as opposed to the automatic function of the expiratory part of the centre, too, is much more marked than that of the inspiratory part.

Taking this in conjunction with the fact that so many of what are known as special respiratory acts, such as coughing, sneezing, and the like, are essentially expiratory in production, it is of evident importance.

Just in the same way as there are nerves, such as the vagus, which when stimulated cause the inspirations to be more rapid and shallower, and finally to cease in inspiration, so there are others which produce the opposite effect, stimulation of which cause the respirations to be deep and slow, and, if the stimulus is prolonged, to cease in expiration; such nerves are the nasal branches of the fifth, the superior and inferior laryngeal, and the cutaneous nerves of the breast and body.

It has been demonstrated by experiment, however, that the nerves which, as a rule, on stimulation produce slowing of respiration may sometimes produce quickening and the reverse. This anomaly is explained by the assumption that different degrees of stimulation of the same nerve may produce the opposite effect, or that the same degree of stimulation may produce the opposite effect according to the different condition, as regards irritability, of the nerve and of the centre.

Although the physiology of the medullary centre and of the nerves connected with it is not completely understood, yet the foregoing considerations enable us to indicate certain ways in which drugs may act in controlling respiratory actions of an abnormal character. Of these drugs we shall mention the several classes when treating of these abnormal actions in the next chapter.

Of the respiratory centre itself, however, we may here mention that its **activity is**:—

  i. **Increased** by heat, the respirations being quicker and deeper. By strychnine, ammonia, atropine, duboisine, brucine, thebaine, apomorphine, emetine, members of the digitalis order, and salts of zinc and copper,
 ii. **First excited and then depressed** by caffeine, colchicine, nicotine, quinine, and saponine.
iii. **Diminished** by cold, the respirations becoming slow and shallow. By chloral, chloroform, ether, alcohol, opium, physostigmine, muscarine, gelsemine, aconite, and veratrine in large doses.

The Heart and Pericardium

The **heart**, enclosed within the bag of the **pericardium**, occupies a considerable portion of the space between the lungs. It is placed in the middle mediastinum, encroaching somewhat on the left lung. Its position on transverse section is shown in Fig. 2, and in relation to the chest wall in Fig. 4.

The **pericardium** on examination is seen to be a

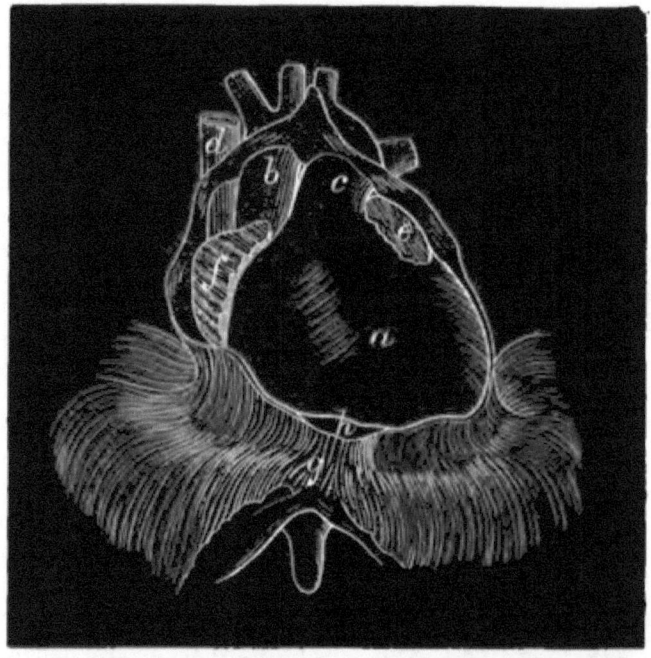

FIG. 6.—The heart within the opened pericardium lying upon the diaphragm. *a*, Right ventricle; *b*, aorta; *c*, pulmonary artery; *d*, vena cava superior; *e*, left auricle; *f*, right auricle; *g*, diaphragm; *h*, cut pericardium. Front view. (After Bourgery.)

more or less conical fibrous bag, from which emerge the great vessels, viz. the aorta and its branches, the superior vena cava, and the pulmonary vessels.

Below it is adherent to the central tendon and adjacent parts of the diaphragm. On opening the pericardium the interior is found to be lined with serous membrane, the visceral layer of which forms a covering for the heart. In the serous sac there is in health a small amount, a drachm or so, of a serous fluid to keep it moist and to moderate friction.

The fibrous pericardium is continued along the great vessels for some distance, being continuous with their external coats; the serous layer internally also covers the great vessels to a greater or less extent and encloses the aorta and pulmonary artery in a common sheath.

The heart and its cavities.—It is scarcely necessary to do more than mention a few of the various anatomical points in connection with the heart. It is free within the sac of the pericardium, its only moorings being the attachment of the latter to the great vessels. It lies with its broadest part or base upwards, backwards, and towards the right, whilst its blunt apex is directed downwards, outwards, and towards the left, striking against the chest wall during life in the left fifth intercostal space, three and a quarter inches from the middle of the sternum bone. The posterior level is from the fifth to the eighth dorsal vertebra. The heart is thus seen to occupy the space behind the sternum and costal cartilage, but is more to the left than to the right. The anterior interventricular furrow divides the heart into two unequal portions, of which the larger is to the right and consists of the right ventricle, whereas the posterior furrow is more to the right than to the left hand. Into the formation of the apex the left ventricle alone enters. There is a deep auriculo-ventricular or transverse furrow. Within these furrows the coronary arteries are situated.

Of the four cavities of the heart the following are the chief anatomical features.

Right auricle.—Is thin walled, quadrilateral in

shape, smooth internally, except in the right auricular appendix, which projects to the left from its anterior and upper angle, where it is ridged with musculi pectinati. It presents at the posterior angles the openings of venæ cavæ (inferior and superior) and besides at the anterior and under part of the cavity, the right auriculo-ventricular opening, which

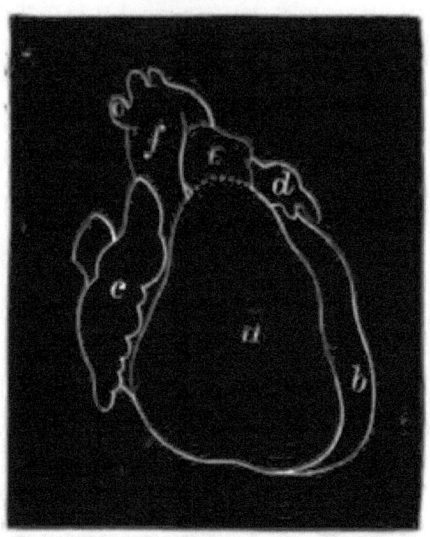

FIG. 7.—The heart as seen from the front. *a*, Right ventricle; *b*, left ventricle; *c*, right auricle; *d*, left auricle; *e*, pulmonary artery; *f*, aorta. (Dalton.)

is oval and large, admitting three fingers. Numerous veins, of which the right coronary is the most important, enter this chamber. This vein, at its termination, is dilated into the coronary sinus, and its entrance into the auricle is guarded by a valve.

Right Ventricle.—The walls of the right ventricle are thicker than those of the right auricle, and its cavity is larger. It forms the chief part of the anterior aspect of the heart, and a small part of the posterior. It does not enter into the formation of the apex. The left angle of the base or upper part of the ventricle, or conus arteriosus, is continued into the pulmonary artery, the entrance of which is guarded by the right semilunar valves. More to the right, is the auriculo-ventricular orifice with its **tricuspid** valve. The interior of the cavity is rough in consequence of the irregular projections of the muscular tissue covered by endocardium forming the so-called *columnæ carneæ*. To some of the columnæ carneæ

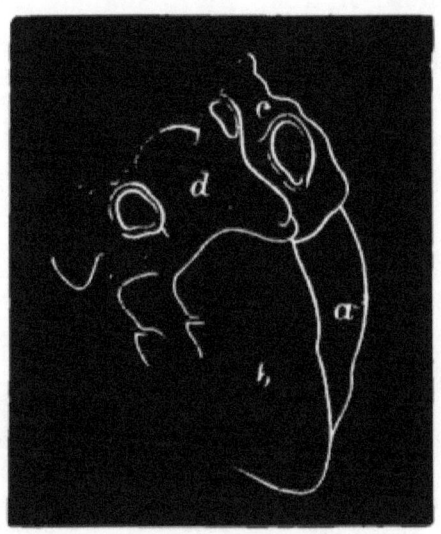

FIG. 8.—The heart as seen from behind. *a*, Right ventricle; *b*, left ventricle; *c*, right auricle; *d*, left auricle. (Dalton.)

are attached tendinous cords (*chordæ tendineæ*) connected with the edges and tips of the segments of the tricuspid valve.

Left Auricle.—This cavity is situated at the back part of the left side of the base of heart. As seen from behind, it is quadrilateral in shape, and has in front its appendix, which is more curved, longer, narrower, and less rough internally than its appendix on the right side. The interior of the cavity is smooth. There are the four non-valvular orifices of the pulmonary veins behind, two on each side. The opening into the left ventricle is situated in the lower and front part of the auricle; it is oval in shape, and slightly smaller than the corresponding orifice on the right side. It is guarded by the mitral or bicuspid valve.

Left Ventricle.—The left is longer and narrower than the right ventricle, as it forms the whole of the apex of the heart. Its cavity, as seen on transverse section, is more or less oval in shape. The internal surface presents very numerous and closely-reticulated columnæ carneæ, smaller than in the right ventricle. The two orifices into the ventricle, viz. the mitral and aortic, are close together at the base, and are situated, the former to the left and behind, the latter to the right and in front. The aortic orifice is circular in shape, and is protected by semilunar valves similar to, but stronger than, the pulmonary valves.

Of the cavities of the heart the ventricles have very much the same capacity during life. The capacity of the auricles is less than that of the ventricles; of the two the right is rather more capacious than the left.

The walls of the left ventricle are two to three times as thick as those of the right—those of the auricles are much thinner.

The whole of the interior of the heart is lined with endocardium, consisting of connective tissue with elastic and some plain muscular fibres, covered with endothelium.

The tricuspid valve is formed of three large and three smaller triangular cusps, made up chiefly of fibrous tissue, covered by two layers of endocardium, the bases of the cusps are attached to the tendinous ring of the orifice; the free edges are attached loosely to the interior of the ventricle by the chordæ tendineæ.

The mitral valve is similarly constructed, but has two large and two small cusps, instead of three as in the other valve.

The semilunar valves on either side consist of three semilunar pouches, like watch-pockets, made up of fibrous tissue covered by endocardium on one side and by the inner coat of the artery on the other. These pouches are attached by their outer and convex border to the wall of the artery, whilst their inner border is free in the interior of the vessel. In the centre of each, at its free edge, is a small fibrous nodule or corpus arantii.

Position of the Lungs and Heart with respect to the Chest Wall

Lungs:
- A. **The right lung.**—
    - i. The **upper lobe** in the front of the chest reaches down to the fourth or fifth rib; laterally to the fourth, and behind to the spine of the scapula.
    - ii. The **lower lobe** behind lies between the spine of the scapula and the tenth rib, and between the eighth and tenth ribs at the side.
    - iii. The **middle lobe** at the side extends between the fourth and sixth ribs and in front of the lower margin of the lungs.
- B. **The left lung.**—
    - i. The **upper lobe** in front extends to the sixth rib in the nipple line, and at the side to the fourth or fifth rib.
    - ii. The **lower lobe** behind extends from the spine of the scapula to the base, and in front and at the sides occupies the space below the upper lobe to the level of the eighth rib.
- C. **The heart.**—
    The **apex-beat** is situated two inches below the left nipple and an inch to the sternal side.

Two thirds of the heart are found on the left side of midsternum.

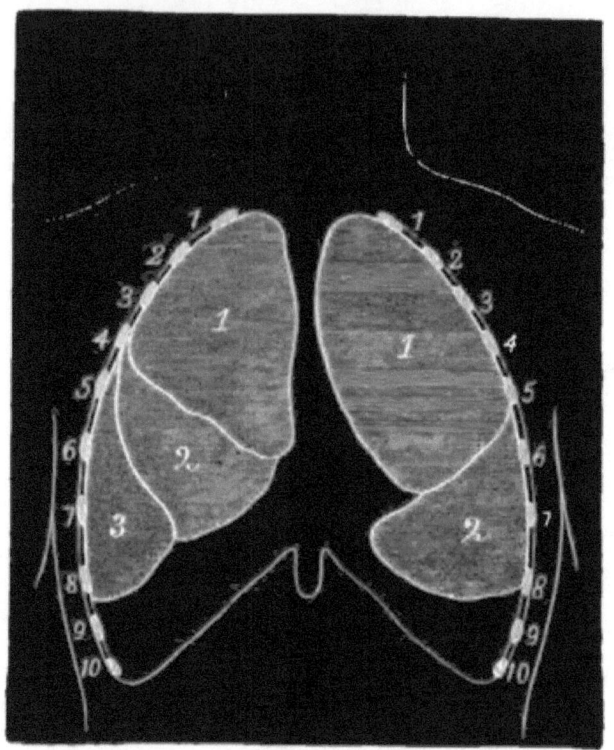

FIG. 9.—Diagram of the chest with the lungs in situ, showing the divisions into lobes (right side, 1, 2, and 3; left, 1 and 2), with the position of the divisions with respect to the chest wall. The limits of the lungs do not of course represent those of the pleura. In the diagram, the shading ought to indicate that the anterior surface of the lungs is rounded and not flat. Front view. (After Bourgery.)

The **right auricle** lies behind the sternal ends of the third, fourth, and fifth costal cartilages and their interspaces, and the edge of the sternum, on the right side.

The **right ventricle** extends from the third to the sixth cartilage on the left side.

The **left auricle** extends from the lower border of the second left cartilage to the upper border of the sternal end of the fourth.

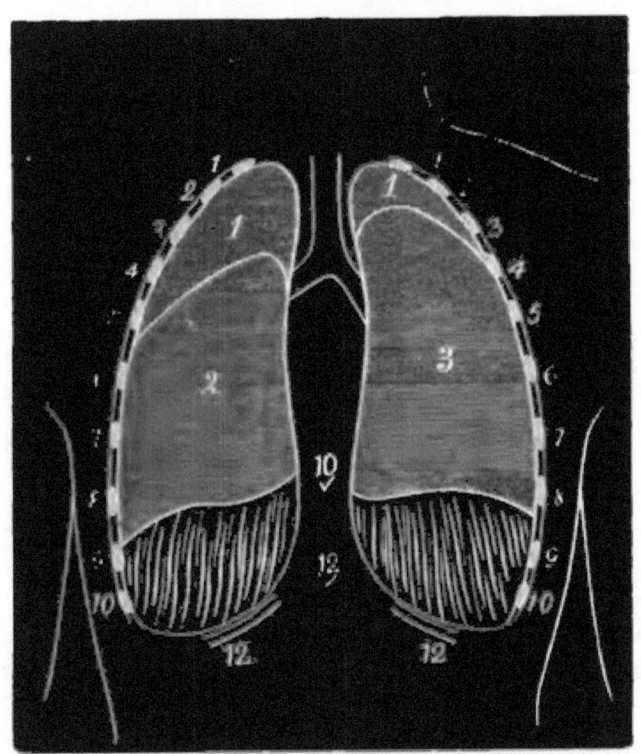

FIG. 10.—Diagram of the lungs within the pleuræ, from behind. On each side the ribs are numbered, in the middle line the position of the tenth and twelfth vertebræ is indicated. A curved line separates the upper (1) from the lowest (3) lobe of the right lung, and a similar line separates the upper (1) from the lower (2) lobes of left lung. (Modified from Bourgery.)

The **auriculo-ventricular sulcus** corresponds with an oblique line from the sternal end of the sixth right costal cartilage to the third left costal cartilage.

Of the valves of the heart.—

The **auriculo-ventricular** valves lie a little to the left of the sulcus, the right are behind the sternum opposite the fourth space, and the left behind the left half of the sternum at the level of the fourth cartilage.

The **pulmonary** valves lie behind the left of the sternum and the third left costal cartilage, the vessel extending to the second cartilage.

The **aortic** valves lie behind left half of the sternum, slightly lower than the pulmonary, opposite the lower part of the third costal cartilage and interspace.

Movements of the Heart.—The exact movements which the human heart executes are chiefly inferred from observation of the hearts of other mammalians. The only other clue which we have to these movements is obtained from the examination of the apex-beat or impulse of the heart against the chest wall. The impulse is due to the contraction or **systole** of the heart. The contraction begins in the auricles, which are filled with blood, at the entrance of the great veins into them, and then spreads downwards to the auriculo-ventricular orifices; the contraction of the ventricles follows, the auriculo-ventricular valves being closed and the semilunars being open. The contraction of the auricles by discharging with considerable force their contents into the already nearly filled ventricles, completes the distention or diastole of the latter and probably stimulates their contraction. This systole of the ventricles discharges the blood in a forcible stream into the large arteries and distends them. This, it will be recollected, causes the pulse and occurs practically at the same time as the apex-beat, and precedes the second sound of the heart.

It is due chiefly to the systole of the ventricles, and the filling of the aorta and pulmonary artery, that the tilting movement of the heart's apex forwards takes place.

The impulse is really the beat of the apex against the chest wall. The base of the heart at the same time becomes less broad and more circular, and approaches the chest wall. The apex of the ventricle is during diastole in contact with the chest wall, and when the ventricle contracts it presses forwards into the intercostal space. The ventricle also twists slightly from left to right. In addition to this, the forcible injection of blood into the aorta and pulmonary artery appears to produce a forward movement or reaction impulse of the heart, which is possibly increased by the tendency of the arteries to elongate. During the pause which follows the ventricular systole, the movements in a reversed direction ensue, and dilatation of the auricles and ventricles takes place. The examination of the characters of the cardiac impulse may be made by means of the cardiograph, an instrument for the invention of which we are indebted to Marey, and to which we shall return in a later chapter, as well as to a consideration of the natural heart-sounds.

Innervation of the heart.—Some of the most important considerations with respect to the heart are those of its nerve-supply and its method of contraction. The chief points about the tissue of the organ itself may readily be recounted. Its muscular tissue occupies, both in structure and action, an intermediate position between skeletal striped muscle on the one hand and involuntary and unstriped muscle on the other. It is a striped muscular tissue, but the striations of its fibres are less marked than in ordinary striped muscle, the fibres are more easily resolved into their elements or muscle-cells; they are branched, and possess no sarcolemma or muscular fibre sheath, and the individual muscle-cells possess a central nucleus. Then as to its action, it is involuntary, but acts in respond to stimuli in a way which resembles yet differs from voluntary muscle. Thus it acts more slowly to the stimuli, contracts more slowly, and although one stimulus will only produce a single contraction, and not a wave of contraction (unless the contraction of the heart be, as some believe, a vermicular contraction), yet a series of stimuli appear to be unable to produce a true tetanus. Of the exact nature of the contraction of the heart-muscle indeed there is still much doubt whether it is a single contraction or twitch, a wave of contraction or a tetanus. Most of the information which we possess upon the subject of the contraction of the heart has been obtained from observations upon hearts of cold-blooded animals, especially of the frog and tortoise, but we must not be too ready to accept without qualification these observations in arguing about the human heart. For example, the heart of a frog will continue to beat, after removal from the body, if placed under favorable conditions for as long as two days and a half. In mammals the heart will beat for some minutes only, although it is true that all power of contraction does not depart for some hours, and a beat may be called forth on stimulation of some kind. As regards the human heart, it is assumed, from observations upon criminals who have been executed, that it approximates in this respect to the hearts of other mammalia.

At any rate we must take it as fairly well established that the beat of the heart is not directly dependent upon the central nervous system. Other observations make it highly probable that the beat depends upon an intrinsic nervous mechanism composed of several groups of ganglion-cells. The situation of these ganglion-cells varies in different animals. In the frog they are chiefly in two positions, viz. in the sinus venosus (Remak's), and in the auriculo-ventricular groove (Bidder's). In mammals the ganglia are found chiefly in the auricles. The groups of ganglion-cells are connected together by nerve-fibres. The ganglia of the sinus venosus and of the auricles are believed to start the movement of contraction, and are looked upon as the dominating motor centres. It was formerly thought that the ventricles in which no ganglia have been found were incapable of independent pulsations, but this has now been disproved. We may therefore suppose that ganglia are not absolutely necessary to ensure the rhythmical pulsation of the heart, but as they more easily respond to stimuli than the muscle itself, they are under ordinary conditions concerned in ordering the rhythmical ventricular contraction.

To insure continuance in order of the contractions of a heart outside the body, it must be supplied with nutritive materials, of which the best is serum or defibrinated blood. It will, however, continue to contract in milk or whey.

Although, as we have seen, the beat of the heart may go on independently of the central nervous system, yet at the same time there is no question but that the heart is in very intimate connection with that system during life, and that its movements may be controlled by efferent influence proceeding from the medulla. The channels of communication by which impulses pass downwards are the cardiac branches of the vagi and sympathetic nerves. If one of the vagi be stimulated, the heart, as a rule, stops in diastole, and if the stimulus be strong, the standstill continues for some time after it has been removed. This is true not only of frogs and mammals but also of man. It has been proved by a personal experiment of Czermak, who was able to press his vagus nerve against a little bony tumour in the neck, and by thus subjecting the nerve to mechanical stimulation was able to slow or even to stop the beating of his own heart. Experiments of various kinds, especially the exhibition of poisons, make it probable that the vagi act by influencing, directly or indirectly, some of the intrinsic ganglia of the heart, and these are therefore called **inhibitory** ganglia. It has been **shown** that in addition to the vagus fibres, **other** nerve-fibres from the cervical and upper thoracic spinal **cord** pass to the **heart,** through the first thoracic ganglion. When these fibres are stimulated, the beat of the heart **is quickened,** although what is gained in rapidity **is lost in strength,** and the blood-pressure is **not increased.** The exact **function of** the **accelerator nerves,** as these **fibres** are called, **is** not properly **understood.** The inhibitory centres **in the** heart can be reflexly stimulated. The efferent **fibres** in the vagi **are** apparently **in** direct **connection with the** intrinsic centres **on the** one hand, and with a **centre** in the medulla, **the cardio-inhibitory, on the other.** The **action of the** cardio-inhibitory centre may be exalted by **afferent** impulses, a stronger stimulus being sent through **the vagi** so as to increase the action of the inhibitory ganglia in the heart, whereby **it beats more slowly** or even stops. This action explains the stopping of **the heart** caused **by** a blow upon the abdomen.

Other **conditions, however, which are not** directly connected with the nervous system, influence the **force and frequency** of the contractions **of the heart,** the principal of which may be **here mentioned.**

> *a*. **A due supply of properly aerated and pure blood.**—Without proper pabulum the heart-muscle is unable to contract properly, **just as a** skeletal muscle would be unable to do its work properly **under** similar conditions. In order that this **shall be supplied in** proper quantity it is essential that **the coronary** vessels should be free and healthy.
>
> *b*. **A proper amount of blood to contract upon.**— The **strength of** the contractions up to a certain point is increased by an increase in the amount of blood contained within **the cavities,** but beyond that point it is diminished, **and if** the cavities be

distended beyond a certain amount the contractions cease.

c. **Increase of blood pressure** increases the resistance to be overcome by the ventricle, and allows an extra supply of blood to the heart substance, through the coronary arteries; it therefore increases the strength of the contraction. It is found also by experiment that the rate of contraction is diminished rather than increased, whenever a rapid pulse accompanies a diminished blood pressure. It should also be recollected that increase of blood in the ventricle increases the blood pressure in two ways, firstly by discharging more blood into the arteries, and secondly by discharging it more forcibly. Thus we see that increased supply of blood raises the blood pressure, and increases the force while diminishing the frequency of the heart-beats.

d. **The outside pressure upon the heart** must not be increased nor diminished beyond a certain point, otherwise the muscle is unable to contract completely, and as a consequence a diminished activity, terminating in complete cessation, results.

Blood pressure.—In the arteries of the body the contained blood under normal conditions exercises a pressure upon their walls which causes their distension, and thus the arteries in health must be considered to be more than full. After each systole of the ventricles this distension is increased, and the pressure of the blood is at this time greatest. The cause of the arterial tension or strain resulting from the pressure of the contained blood upon the walls of the vessels is threefold: (1) That the arterial walls are elastic and distensible. (2) That the entrance of blood into the arteries is accomplished by the forcible contraction of the heart. (3) That the exit of blood from the arteries is obstructed by having to traverse the minute channels of the small arteries or arterioles and capillaries. In other words, there are three principal factors in maintaining arterial blood pressure,—the heart, the arterial walls, and the peripheral resistance. It has been shown that a fall of arterial **tension** below a certain point is incompatible with **life**, and no doubt a healthy condition is only **maintained as long as** the blood pressure is kept within certain and **narrow** limits. It is evident, therefore, that the conditions for maintaining the blood pressure at its **normal** point are very important.

The state of fulness of the arteries depends upon **(1) the amount of blood which they contain at the beginning of the systole of the heart,** and (2) **the amount of blood pumped into them by the systole**.

(1) The amount of blood in the arteries at the beginning of the systole evidently depends upon the rapidity with which the **blood** can flow into the veins **and the frequency of** the contractions **of** the heart. It is evident that if the arterioles and capillaries are dilated, more blood will flow through them in a given time than if they are contracted, and also **that if there is more time between** the beats of the heart, more **blood** will pass on into the veins during the intervals. Under such conditions, therefore, dilated arterioles and capillaries and slowing of the heart's beats would diminish arterial fulness, or in other words would diminish the blood pressure, and similarly, without considering other conditions, contracted arterioles and capillaries, or increased peripheral resistance, as it is **aptly** termed, would have the opposite effect.

(2) The heart may tend to fill the arteries or to **raise the** blood pressure by an increase of the force **or of the** frequency of its beats, or by discharging an increased amount of blood at each systole and the reverse.

To summarise some of the previous considerations about these three factors:—

(*a*) The **force** of the contractions of the heart

depends upon the amount and nutritive activity of the muscular fibres, as well as upon other circumstances which are much the same as those concerned in the contraction of a skeletal muscle. Thus it is obvious that, *cæteris paribus*, a larger heart will contract with greater force than a smaller, and that a healthy fibre will contract with greater power than a degenerated one. But the conditions necessary for due nutrition are not simple. One thing at any rate is undoubtedly essential, *i. e.* a **due supply of healthy and properly oxygenated blood**. This we have laid stress upon before. Another thing is **connection with the nervous system**, and particularly **a healthy condition of the intrinsic motor ganglia**, a muscle being much more irritable to stimuli applied through the nerves. Again, in order that a muscle shall contract to the best advantage it must be properly weighted, but not too heavily, and so in order that the heart-muscle may do likewise, it must have a certain amount of resistance to overcome; it must contract upon **blood sufficient to fill the cavities** but not to distend them beyond a certain point, **the resistance to the exit of blood** from a cavity **must be neither too great nor yet too small**; and lastly, the fibres must not be made to **contract too often**, nor yet too seldom, or the result will be exhaustion from **overuse on the one hand, or debility from under-use on the other.**

(*b*) The **frequency** of the contractions of the heart depends chiefly upon the action of the vagus in stimulating the inhibitory ganglia or rather of the vagus (cardio-inhibitory) **centre which**, as has been described, may be stimulated, either directly or reflexly, to send impulses down the vagus nerves. The usual method of stimulation is an increased blood pressure.

As an increase of the contraction of the small arteries may cause a slowing of the heart, so too in the heart there is a nerve, the **depressor**, which may produce, by acting upon the medullary vaso-motor centre, relaxation of the small arteries and diminished blood pressure. The method of stimulation of the depressor is possibly an increase of endocardial pressure beyond a certain amount.

Another cause, whether mediate or immediate is uncertain, which produces an increase in the frequency of the heart's beats, is a diminution of the amount of blood contained in its cavities,—incomplete dilatation producing incomplete contraction.

We know little of the action of the accelerator nerves in this matter. It appears that they are not antagonistic to the vagi, but whether they act at all under ordinary conditions is a moot point.

The intrinsic ganglia, again, are, as far as we know, of two, if not three, kinds. Their relations to one another are not understood,—whether they are alternately in action, and, so to speak, complementary, or whether they are independent, cannot be said to be determined. (*a*) Motor ganglia which under ordinary circumstances originate the rhythmical contractions of the heart. (*b*) Inhibitory ganglia, through which the vagus acts directly or indirectly; and (*c*) Accelerating ganglia, through which the accelerator fibres act, but about this little is absolutely certain.

(*c*) The **amount of blood discharged** by each beat of the heart depends upon:—i. The **completeness of its systole**, and this again depends upon the **completeness of its diastole, for it is evident** that if only a little blood reaches it from the lungs that the systole must be incomplete. A diminished amount will flow from the pulmonary vessels into the heart either if there is pulmonary obstruction of any kind or if little blood pass to the lungs from the right side of the heart from over-dilatation of the systemic vessels, *e. g.* from loss of the vaso-motor control over their calibre.

ii. The total amount of blood in the **body.** This, however, strangely enough, is almost inoperative, as it requires very copious bleeding to diminish, and the injection of a very large amount of blood into the vessels of an animal to increase, the blood pressure.

Thus we see **blood pressure is increased** by:

(*a*) A more rapid or a more forcible or complete contraction.

(*b*) An increase of the peripheral resistance.

(*c*) An increase in the total amount of blood in the body.

And **blood pressure is diminished** by:

(*a*) A less rapid or less forcible or complete contraction.

(*b*) A diminution of the peripheral resistance.

(*c*) A diminution in the total amount of blood in the body.

Let us consider next, in the briefest possible manner, the **peripheral resistance.** The conditions which affect this, omitting for one moment the organic condition of the vessels themselves, are almost exclusively connected with the nervous system. By it the calibre of the vessels is regulated. There are, it will be remembered, no less than three sets of centres which take part in the control, viz. the medullary centre, the spinal centres, and the local centres about the vessels themselves. Whereas the last can be stimulated only directly by an alteration in the blood, the others are capable besides of being reflexly affected.

Thus impressions carried to the medullary centre may reflexly cause dilatation or further contraction of the arteries, and so with the other centres.

Lastly, apart from their nerve-supply, the condition of the coats of the large arteries, which affects the blood pressure, is the amount of their distensibility. When they become more or less rigid from old age or disease variations of blood pressure naturally become less and less marked.

As regards the drugs which act upon the heart or upon the peripheral resistance respectively, there are, according to Brunton, these six varieties, viz.:

1. **Cardiac stimulants,** *i. e.* "substances which rapidly increase the force and frequency of the pulse in conditions of depression. The most important are ammonia and alcohol in its various forms. . ."

2. **Vascular stimulants,** *i. e.* "substances which cause dilatation of the peripheral vessels, and thus render the flow of blood through them more rapid. The most important are heat, alcohol in its various forms, ether, nitrous ether, Dover's powder, and acetate of ammonia."

3. **Cardiac tonics.**—"Drugs which have no perceptible immediate action on the heart, but when given for a little while render its beats much more powerful, although usually much slower. The most important are digitalis, digitalin, digitalein, digitoxin, erythrophlœum (casca), erythrophlæin, strophanthus hispidus, strophanthine, convallaria majalis, convallamarin, adonis vernalis, adonidin, squills, scillain, helleborein, antiarin, caffeine, nux vomica, strychnine."

4. **Vascular tonics.**—"Substances which cause increased contraction of the arterioles or capillaries. They not only raise the blood pressure but influence to a considerable extent the quantity of lymph poured out into the tissues or absorbed from them, and thus modify tissue change. They are of special importance in the treatment of dropsy. The most important are digitalis, iron, and strychnine."

5. **Cardiac sedatives.**—"Substances which lessen the force and frequency of the heart's action." The chief are aconite, veratrum viride, antimonial preparations.

6. **Vascular sedatives.**—"Remedies which by increasing the contraction of the vessels lessen the flow of blood through them." They are cold, digitalis, ergot, hamamelis, lead acetate, opium.

Other Thoracic Viscera

Of the other viscera within the chest it will be only necessary to say a few words about the **aorta**, for a study of its relations to other structures in the chest indicates to a certain extent the position of the latter.

The aorta within the thorax is arbitrarily divided into two parts, (*a*) **the arch**, which extends from its origin at the left ventricle and passes up to the right side on a level with the lower border of the third left costal cartilage, and slightly forwards to the second right costal cartilage; it then goes backwards to the left side of the body of the second dorsal vertebra, at first ascending slightly, passes downwards, bending a little towards the middle line, and descends upon the left side of the body of the third dorsal vertebra; becomes (*b*) the **thoracic aorta**, and passes downwards upon the bodies of the dorsal vertebræ, getting gradually towards the middle line.

The relations of the aorta with regard to the various structures in the neighbourhood are important with reference to the symptoms produced in aneurysm of the various parts, ascending, transverse, or descending, of the arch of the vessel, and perhaps it will not be wasted space to mention them briefly.

(1) **Ascending.**—This portion, two and three quarter inches long, is contained almost entirely within the pericardium in the same sheath and behind the pulmonary artery. It then has in front the pulmonary artery and the right auricular appendix, as it passes to the right, it has the pulmonary artery to the left, and the sternum, connective tissue, and remains of the thyroid gland in front, the descending cava to the right, the pulmonary vessels and the right bronchus behind.

(2) **Transverse.**—Has covering it on the left, the left pleura and lung. Behind it and to the right are the trachea, œsophagus, and thoracic duct. It gives

off above, main branches, to the head and upper extremities, in front of which passes almost transversely the left innominate vein. It also has in front of it the left vagus and cardiac nerves, and the left recurrent laryngeal is hid beneath and behind it. Below it is the pulmonary artery dividing into two branches. It is connected with the left branch of this vessel by means of the ductus arteriosus.

(3) **Descending.**—This part lies upon the body of the third dorsal vertebra, and has in front of it the pleura and root of the left lung. To the right lie the œsophagus with the thoracic duct.

The thoracic **aorta** is situated in the posterior mediastinum (Fig. 2), lies upon the vertebral bodies, and has in front of it the root of the left lung and the pericardium; on the left lies the left pleura.

CHAPTER II

THE SYMPTOMS OF THE DISEASES OF THE CHEST

It is not surprising, when we consider the importance of the thoracic viscera, the heart and the lungs forming two of the three organs, the brain being the third, which have been called the **tripod of life**, that the derangement of their functions, which must necessarily, to a greater or lesser extent, occur when they become the subject of diseased processes, should give rise to distinct symptoms, many of which are of considerable gravity.

The value of such symptoms, when present, in the diagnosis of diseases of the chest is, as may be imagined, very great, and can hardly be over-estimated. It is necessary, however, to remember that many of them, although so constantly associated with chest diseases as to be called *par excellence* **chest symptoms** may yet occasionally arise under other conditions in which the chest is affected not directly if at at all. In order to guard ourselves against this possible source of fallacy let us now consider somewhat in detail the most important of these symptoms and their possible causes.

We will treat of them in the following order:
  i. **Hyperpnœa, Dyspnœa, and Asphyxia.**
  ii. **Cough and Sputum.**
  iii. **Hæmoptysis.**
  iv. **Palpitation of the Heart.**
  v. **Angina Pectoris.**
  vi. **Syncope.**

I. Hyperpnœa, Dyspnœa and Asphyxia

Of all the symptoms indicating some disturbance of the respiratory functions, **rapidity of breathing or hyperpnœa** and **difficulty in breathing, or dyspnœa**, are among the most constant. Both of these symptoms, however, may occur under conditions which do not primarily affect the lungs. This will be at once understood if reference be made to what has already been said as to the action of the medullary and spinal respiratory centres in health (p. 23).

It has been demonstrated that with regard to these centres the condition which maintains a given rate of respirations per minute under ordinary circumstances is that the gaseous interchange brought about by each respiration should be maintained at a definite standard. If this standard be lowered the rapidity of respiration must increase; if the standard be raised, as by an extraordinary inspiration, the respirations become slower.

Many causes may produce an alteration of the natural interchange of gases in the pulmonary alveoli. Some of these are simple and straightforward, as, for example, an obstruction to the due entry of air into the lungs, such obstruction taking place in the larger or in the smaller tubes. Again, air may enter in sufficient amount, but not containing a due amount of oxygen, or again, the blood may not be in the condition to take up a requisite supply of oxygen. But besides these causes there are others more remote, namely, the conditions of the system in which the absorbed oxygen is more rapidly used up in the nutrition of the tissues than is normally the case, whereby a most abundant supply of this gas is required, and is to be obtained only by more active respiration.

As long as by increased rapidity of breathing the supply of oxygen to the blood, and so to the respiratory centres, can be kept up to the demand, the condition may be called hyperpnœa only. But this state seldom or never lasts for any length of time, but passes into dyspnœa, or **difficulty of breathing**, or what is called by patients **shortness of breath**, This arises as soon as rapidity of breathing alone is unable to supply the oxygen wanted, the tension of that gas in the blood becomes less and less, and the medullary centre is more and more stimulated to action, and sends out motor impulses not only more rapidly but also to a wider area, so that the muscles of extraordinary inspiration are stimulated to action. By this action it may be that the area of the chest is increased, and more air is supplied to the lungs; still, however, respiration is difficult and dyspnœa persists. Sometimes it happens that the respiration is almost impossible except when the patient is sitting or standing up. This modification of dyspnœa is known as **orthopnœa**.

Dyspnœa passes into **asphyxia** when, in spite of the extraordinary effects of respiration and their increased rapidity, the oxygen standard becomes lowered beyond a certain point. When asphyxia commences the respiratory centre sends out more and more energetic stimuli to all the inspiratory muscles, and violent contractions ensue with marked action also of the expiratory muscles following. Next, convulsions of the muscles of the whole of the body may occur. After a varying time these symptoms of irritation pass into the symptoms of paralysis, and slow but deep and quiet respirations ensue. These last a varying time and then cease.

During the first two stages the blood pressure rises, and the pulse is quick, free, and firm, but afterwards there is a rapid fall in pressure below the natural, but the heart may continue to beat for some seconds after respirations have ceased.

After death the venous system and the right side of the heart are found to be full of blood, the arteries and left side of the heart being as a general rule empty. It should be remembered that each stage of asphyxia varies in duration according as the deprivation of air is a sudden or gradual process. For example, an animal such as a dog, deprived of air suddenly by blocking of the trachea, passes through the stages to death in about five minutes, whilst in a patient suffering from some chronic lung disease the early stages may be very prolonged, lasting for weeks or months.

The **condition of the vascular system** is a factor in gradual as well as in more rapid asphyxia, which soon complicates the situation and must not be overlooked; as the blood becomes more and more venous in character, the peripheral and central vaso-motor centres are stimulated to greater and greater action, whereby the arteries of the body are contracted and the blood pressure is raised, the veins are gorged with blood, and the heart tends to become distended. The veins if opened for a time act like arteries, and send out the blood in a jetting stream. Thus it is that in impending asphyxia, temporary relief may be obtained and the strength of the heart's beat may be increased by abstraction of blood from the veins.

The very **gradual suffocation** of patients whose lungs are little by little rendered impervious to air by phthisis and other diseases is no doubt due to the fact that more active respiration is carried on by what remains of the lung tissue, and also that the hæmoglobin exists in the venous blood under these conditions chiefly as reduced hæmoglobin, and is therefore just in that state to seize upon what little oxygen it meets with in the alveoli. It is the special and natural arrangement that hæmoglobin can take up almost its greatest possible amount of oxygen from what may be considered to be a very diluted volume of the gas.

So far nothing has been said of another factor in the condition of gradual suffocation which is the frequent termination of so many chest affections, that is to say, of the gradual accumulation of carbonic acid gas in the blood. As with the diminution of the oxygen standard, so the increase of carbonic acid gas is a very gradual process but yet it is undoubtedly one which is constantly going on. My own belief certainly is that the mode of death in chest disease is not simply one of asphyxia. There is, to my mind, strong evidence that that gradual drowsiness and lethargy passing into coma which one so often observes as the final scene in such cases, owes its origin to the non-elimination of carbonic acid gas and other effete products and their gradual accumulation in the blood.

We may say then that dyspnœa, including in its first stage hyperpnœa, is due to any cause which interferes with the due oxygen supply to the respiratory centres in a given time. These causes may be considered as they are tabulated (p. 52) under three heads.

    A. **Respiratory dyspnœa.**—When there is either some defect in the respiratory apparatus, thereby preventing the air from coming in contact with the blood circulating in the alveolar vessels, either due to deficiency in amount of air or abnormality in its quality.

    B. **Hæmic dyspnœa.**—Where the fault lies with the quantity of the blood or the amount circulating within the alveolar vessels in a given time.

    C. **Centric dyspnœa.**—Where the dyspnœa arises from some special action of the respiratory centre itself, whereby it is locally influenced. The only example of this about which we know anything, is the so-called *heat* dyspnœa arising in fevers. The raised temperature of the blood increases the nutritive activity of the centre.

Table of the Causes of Dyspnœa.

A. RESPIRATORY.

- Air* reaches the blood
  - **Deficient in quantity, e.g. from**
    - Diminished capacity of the thorax due to
      - Contractions.
      - Malformations.
      - Tumours ⎱ Thoracic or abdominal.
      - Effusions ⎰
    - Imperfection of respiratory movements due to
      - Pain, e.g. in pleurisy or peritonitis.
      - Paralysis of motor nerves, e.g. phrenic.
      - Muscular weakness, e.g. degeneration or wasting of muscles.
      - Adhesions, as in chronic pleurisy.
      - Too yielding walls, as in rickets.
      - Loss of elasticity of thoracic walls or lungs.
    - Diminished calibre of air passages from causes
      - Extramural, e.g. tumours pressing on any part, as enlarged thyroid, aneurysm, &c.
      - Intramural—Spasm of larynx or bronchi, swelling of any part of respiratory tract, from mouth and nares downwards, e.g. œdema of larynx, malignant growths, &c.
      - Intermural—Foreign bodies, false membranes, accumulated secretions.
    - Affections of parenchyma of lungs, e.g.
      - New growths, e.g. tubercle, carcinoma.
      - Pneumonia.
      - Fibroid induration.
      - Emphysema.
  - **Deficient in quality, e.g. from**
    - Rarefaction, e.g. from
      - Heat.
      - Diminished atmospheric pressure.
    - Noxious gases, e.g. chlorine.
    - Fevers, &c.

CAUSES OF DYSPNŒA

B. HÆMIC. *Blood reaches the air altered in*

- **Quantity**
  - In excess, *e.g.* effects of exercise
  - In defect, *e.g.* After hæmorrhage.
  - In interference with pulmonary circulation by, *e.g.*
    - Pericarditis.
    - Morbus cordis.
    - Embolism of pulmonary artery.
    - Emphysema.
    - Fibroid induration.
    - Arterial spasm.
- **Quality**—*e.g.* in chlorosis, fevers, poisoning by carbonic oxide gas.

C. CENTRIC. *Blood reaches the centre*—abnormal in quantity or quality, in cases as in B; but the effect of too rapid circulation and increased temperature of the blood, as in fevers, produces special form of dyspnœa, viz. *Heat dyspnœa*.

---

* *Note.*—Although an increase of oxygen in the air breathed up to a certain point produces not dyspnœa but *apnœa*—a very great increase of the oxygen tension, *i.e.* of oxygen about 20 times the normal, produces in animals all the symptoms of asphyxia.

II. Cough

Another of the most constant symptoms of chest disease is **cough**. The act of coughing consists in a series of co-ordinated muscular movements, viz. of an inspiration deeper than usual, next a firm closure of the glottis, followed by one or more short but violent expiratory efforts, whereby the air is forced through the glottis and the characteristic sound is produced.

The act may be **voluntary or involuntary** and **reflex**. The cause is the presence of something which irritates the **sensory nerve terminations**. The situation of the irritation and its nature allow of the classification of coughs in the following manner:

1. Coughs due **to the irritation of the fibres of the vagus nerve or of its branches**, especially of the superior laryngeal, which supplies the whole of the mucous membrane of the larynx, or in the mucous membrane of the pharynx, larynx, trachea, or bronchi, in the pleuræ, or lung.

   a. By congestion, inflammation, ulceration, new growths, or other morbid conditions, whereby a peculiar sensitiveness is developed or the sensory nerve terminations are directly pressed up. The laryngeal mucous membrane is by far the most sensitive naturally, and the sensitiveness diminishes downwards; this is the case also in disease.

   b. By the inhalation of cold air or of irritating gases, *e.g.* chlorine, or of dust, food, coins, or other foreign matters. As these irritations will act when the mucous membrane is normal, they are likely to act more powerfully, in diseased states, unless these are very chronic.

   c. By the presence of blood or of secretion upon the mucous surface, *e.g.* mucus, pus, or false membrane.

This class includes by far the most common varieties of cough.

2. Coughs due to **the irritation of the fibres of the vagus, and of other** sensory **nerves distributed elsewhere,** *e.g.* in the external auditory meatus, nose, and œsophagus, liver, and spleen, heart and pericardium, pharynx (glosso-pharyngeal), stomach and alimentary canal (**stomach**-cough), uterus, &c., and external surface of the body.

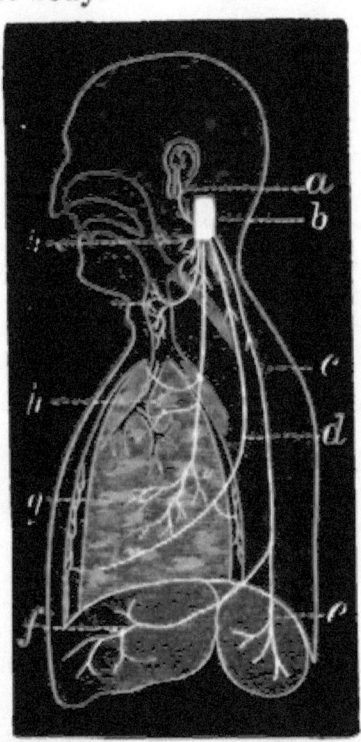

Fig. 11.—Diagram of the afferent nerves which when irritated stimulate the respiratory centres (*b*) and produce **coughing.** The nerves are distributed—*a,* to the ext. auditory meatus; upper *h,* to the pharynx; *i,* to the larynx; lower *h,* to the bronchi; *g,* to the lung; *d,* to the pleura; *f,* to the liver; *c,* to the spleen and œsophagus. (Brunton.)

3. Coughs due to **the irritation by direct pressure upon the trunk of the vagus or superior laryngeal nerve,** *e.g.* by aneurysms and other tumours.

4. Cough due to **centric causes.**

 a. By circulation of unhealthy blood through the nerve-centre which presides over the co-ordinated action. This, it is suggested, is what occurs in gout and allied conditions.

 b. Organic or functional disease of the brain.

Character.—Few coughs are very characteristic of special chest affections.

Their **Tone** varies with the condition of the larynx, being **loud and hollow** when the larynx is almost healthy and the cords vibrate well; hoarse, **wheezy or voiceless** when the larynx is much obstructed with chronic induration, or with mucous or other deposit; **metallic or brassy** in croup. It is **whooping or crowing** in whooping-cough; **whistling** in spasmodic croup; **wheezing** in bronchial catarrh; **paroxysmal,** chiefly in whooping-cough, spasmodic croup, obstruction in the trachea or bronchi, and in acute catarrh of the mucous membrane; these are especially involuntary and uncontrollable.

Nearly all coughs are preceded by a tickling in the throat, and although many may be called reflex, yet in nearly every case the irritation which sets up coughing is perceived by the brain.

The character of the cough, according to Brunton, depends also upon the seat of the irritation; if it be where expiratory nerves are chiefly distributed, as in the larynx, trachea, and bronchi, and pharynx, it is **loud, explosive, and prolonged**; if the inspiratory nerves be affected for the most part, **short and hacking.** If the irritation be in the pharynx, the cough is frequently attended by vomiting.

Coughs are called **dry** or **moist** according as the expectoration is absent or abundant. When expectoration is present, its amount, characters, composition and appearance are of considerable value in diagnosis.

 Sputum.—The expectoration may differ in:

  1. Composition.—It may consist simply of (*a*) mucus resembling the natural secretion exag-

gerated, or frothy, as in acute catarrh of the respiratory mucous membrane; (*b*) **muco-purulent.** In this the mucus is mixed with pus more or less intimately. This kind of sputum is common to nearly all affections of the respiratory mucous membrane. The pus-cells are seen under the microscope to form a considerable proportion of the material. (*c*) **purulent.**—The expectoration of pure pus is rare. When it occurs it is generally in cases of cavities secreting pus or in empyemas discharging through the lung; sometimes the sputum of chronic bronchitis practically consists of pus only.

2. **Colour.**—The sputum may be **whitish or colourless,** in pure mucous expectoration.

Yellowish in muco-purulent expectoration.

Greenish in purulent expectoration.

Yellowish red in bronchiectasis.

Yellow in jaundice, and in hepatic abscess opening through the lung.

Black.—Either coloured by presence of carbon particles free or contained in epithelial cells. It occurs in foggy weather, and in occupations in which there is much dust, which may set up phthisis as well as chronic bronchitis.

The four following arise from the presence of blood, viz.:

Rusty, in acute pneumonia from a slight admixture of blood.

Prune juice, in acute pneumonia from a greater admixture of blood.

Blood stained, varying from specks or streaks mixed with mucus, to pure unchanged blood.

Changed blood may give the sputum a green or yellow colour. In very warm weather various colours may develop in sputum which has been allowed to stand for some time, such as **egg-yolk and grass green**. These colours appear to arise from the **growth** of fungi, and **are** clinical curiosities.

3. **Consistence.**—The sputum of croupous pneumonia being excessively viscid; the frothy **sputum of** acute bronchial catarrh coming **next, and the purulent** sputum has the least consistency.

4. **Odour.**—The sputum in **cases of** dilated bronchi may be rank and **fœtid,** and in **gangrene** it is always extremely fœtid; that **from** a phthisical cavity too is sometimes also of disagreeable and almost characteristic **odour.**

In the expectoration, besides **the** usual constituents, of which the **mucin** and the **albumen** are generally much increased, we may have **columnar ciliated epithelium** (rarely), **pus-corpuscles, blood-corpuscles, debris of lung tissue,** especially elastic fibres* (phthisis pulmonalis), **fibrinous casts** of the bronchial tubes (in croupous bronchitis), **crystals of the fatty acids** (in bronchiectasis and gangrene especially), **and Charcot's crystals,** elongated very pointed octohedra, rhombic plates, or sharp and spindle-shaped bodies of uncertain nature (in asthma and other bronchial affections). **Microorganisms** of various kinds (micrococci, **bacteria,** bacilli and vibrios), **and** accidental constituents.

Methods of examination of the sputum.—In addition to the naked-eye characters, sputum should be also examined with the aid of the microscope.

i. The simplest method is to place a small droplet upon a large cover glass, and press it down between the cover glass and slide. By this means a sufficiently thin layer is obtained, which should be examined with $\frac{1}{4}$ or $\frac{1}{6}$ inch objective, and many of the above-mentioned substances may be distinguished if present.

ii. Another and more effectual method is to dry a thin film upon a cover glass in the flame of a spirit lamp, and to double stain it, if tubercle bacilli be suspected with the

* See under Phthisis.

tubercle stains as indicated below, or if not, first with an aqueous solution of methylene blue for two minutes, and after washing in water with an alcoholic solution of eosin for two minutes, rewashing, drying in the flame of a spirit lamp, and mounting in Canada balsam.

iii. To find elastic fibres in sputum in cases of phthisis, it is best to collect the sputum for twenty-four hours, boil it for some long time with rather more than its own bulk of caustic potash, and then to pour it into a large quantity of distilled water and examine the sediment under the microscope.

iv. To stain sputum for tubercle bacilli, the following method, of the certainty of which from much experience we can vouch, is strongly recommended:

A small droplet of sputum is taken up with a not too fine capillary pipette, and is deposited on the centre of a square cover glass ($\frac{7}{8}$ inch). Another cover glass is now placed upon the sputum, and the two are squeezed together and then separated by sliding one away from the other. A small quantity of sputum then adheres to both cover glasses in a thin layer. They should then be dried in the flame of a spirit lamp. They are now ready for staining. The first stain is fuchsine solution; it should be prepared thus:

Saturated watery solution of **aniline oil**	100 cc.
Saturated alcoholic solution of **fuchsin**	15 cc.
Absolute alcohol	10 cc.

Half fill an ordinary-sized test tube with this solution, warm the solution to 45° C., filter into a porcelain dish, and place the cover glasses sputum downwards upon the surface of this solution, and retain them in this solution for ten minutes, which should be kept at same temperature; then transfer into the washing fluid thus made:

Of a 10 per cent. solution of nitric acid	1 part
Methylated spirit	1 part
Distilled water	1 part

and wash thoroughly for several minutes.

Then transfer to a filtered (saturated) solution of methylene blue for five to ten minutes. Afterwards wash in a saturated solution of sodium carbonate for fifteen minutes, then in water.

Next allow the water to drain off the cover glasses, dry in the flame of the spirit lamp, and mount in Canada balsam dissolved in benzol.

Although a long process, by its use the crystals so often present in the nitric-acid-washed specimens are avoided, and the methylene blue forms an excellent background for the red bacilli.

It is unnecessary to mention other methods of staining the

bacilli. They are all more or less similar, and modifications of Ehrlich's formula.

Treatment.—It is only requisite in this chapter to indicate the general plan of treatment of coughs, and may be summarised in the following propositions:

1. Seek for and if possible **remove the irritation** producing it, *e.g.* foreign bodies, such as growths, &c., from larynx; diminish congestion; relieve constipation, teething, or worms; treat uterine affections.
2. **Allay irritation**, in different ways according to its position; gargles for enlarged tonsils, swollen uvula and relaxed throat; inhalations of steam, with or without antispasmodic drugs, or of creosote, benzoin or carbolic acid; demulcent soothing drinks, linseed tea, liquorice, jellies, either alone or mixed with mild opiates, often allay irritation of the throat and pharynx.
3. **Help expectoration** sometimes by increasing secretion, or by stimulating expectorants.

Expectorants

Expectorants are drugs which are employed to get rid of excess of secretion from the air passages. They act in various ways, either upon the expulsive mechanism, rendering the efforts of the expiratory muscles more powerful, or upon the secretion, lessening it perhaps or rendering it less adhesive and more copious, but more capable of being easily expelled.

CLASS I.—**Drugs which increase the activity of the respiratory centres.**—Strychnine, ammonia, emetine, ipecacuanha, belladonna, atropine, senega, and saponine.

CLASS II.—**Depressing expectorants**, *i.e.* drugs which increase secretion, generally tending to lessen blood pressure. Such are antimonial preparations, tartar emetic, alkalies, ipecacuanha, emetine, lobelia, lobeline, jaborandi, pilocarpine, apomorphine, Quebracho, aspidospermine, potassium iodide.

CLASS III.—**Stimulating expectorants**, *i.e.* drugs which diminish secretion and increase blood-pressure. Such are acids, ammonium chloride, ammonia, strychnine, nux vomica, senega (saponine), **onion, garlic**, squill, tar, benzoin, benzoic acid, balsams of tolu and Peru, wood tar, turpentine, oleum pini Sylvestris, oleum pini pumilionis, sulphur, saccharine substances, liquorice.

4. **Diminish secretion**, when very excessive, *e.g.* by opium, counter-irritants, *e.g.* blisters to the surface; or by atropine, which is said to be more powerful than opium.
5. **Administer** special drugs when the cause of the cough is specific, *e.g.* iodide of potassium, belladonna, chloroform, and the like.
6. **Emetics** when nothing else will cause expectoration of irritating materials.
7. **Diminish the excitability of the respiratory centres** so that they may respond less energetically to stimuli (see p. 25).

III. Hæmoptysis

Hæmoptysis consists in the expectoration of blood from the air passages. The blood is generally the result of **capillary** hæmorrhage, but may occasionally arise from the rupture of a **larger vessel**, or of a **small aneurysm** of one of the branches of the pulmonary artery. Leaking of an **aortic aneurysm** into the trachea also occurs every now and then.

The causes of the **hæmorrhage** are as follows:
1. Congestion, active or arterial (flexional hyperæmia), and passive or venous (mechanical hyperæmia), inflammation, ulceration, or other morbid conditions of the **larynx, trachea, or bronchi** themselves.

2. Congestion, active or passive, inflammation, consolidation, syphilis, cancer, and other **conditions of the lung-substance.**
3. **Mediastinal tumours, enlarged bronchial glands,** pressing upon and causing erosion of the bronchi.
4. **Pressure upon and leaking of an aneurysm** into some **part of the respiratory** tract, especially the **bronchi or trachea.**
5. **Heart** or **vascular disease,** especially mitral **valve** disease or dilated right heart.
6. Peculiar delicacy of the mucous membrane.
7. Purpura, scurvy **or** hæmophilia.
8. Blood from nose or pharynx trickling **into** larynx.
9. **Mechanical causes,** *e. g.* straining, blowing upon a wind **instrument, coughing, foreign** bodies **or** irritating substance.

Characteristics **of** blood-spitting.—(*a*) **Of the** attack.—Generally sudden, with or without a sensation of **weight in** the **chest,** brought on by **coughing** if in moderate amount, or without exertion in **gulps,** or sometimes apparently **vomited** if in large quantities. Often accompanied after the first day by more **or** less **fever.** Variable in quantity from a mere streak or two to a pint in the course of several days. The blood as a rule gradually diminishes after the first day of the attack. (*b*) **Of the blood.**—Bright, frothy, fluid, mixed with phlegm at some period, alkaline in reaction, salt to the taste.

Diagnosis.—Bleeding from the nose (epistaxis) or from the pharynx, mouth, or gums may be mistaken for true hæmoptysis; care must be taken therefore in the examination of the localities from which the blood may have come, and also of the chest; this, coupled with the history of the attack, its onset and duration, and the quantity and quality of the blood, will as a rule determine its source.

Hæmoptysis may be mistaken for hæmorrhage from the stomach (hæmatemesis).

Hæmoptysis *v.* Hæmatemesis

Hæmoptysis.	Hæmatemesis.
Kind of blood.	
Bright red, frothy fluid, and alkaline; salt to the taste.	Dark purple, solid in clots, acid.
Quantity.	
Variable, usually less in amount.	Very large.
How brought up.	
Usually by coughing.	By vomiting.
Accompanying circumstances.	
Frequent symptoms of chest disease, râles in chest.	Anæmia, dyspepsia, pain in region of stomach, vomiting; or ascites, hæmorrhoids, and other symptoms of portal obstruction before the attack occurs.
Sequelæ.	
Blood continues in sputum more or less for several days, or a week. There is very frequently fever. T. 100—102° F.	Melænæ. The vomiting of the blood is often preceded by fainting.

Prognosis.—Death from hæmoptysis is comparatively rare. When it does occur it is generally due to rupture of one of the aneurysms before mentioned. If the hæmoptysis is due to pressure of an aortic aneurysm upon and leaking of it into the trachea death is certainly imminent. The connection between hæmoptysis and phthisis will be treated of under the latter head.

Treatment.—The patient should be kept quiet in bed in a cool, well-ventilated room, should be fed on slop diet, milk and arrowroot principally; should be given ice to suck, and his bowels should be freely opened by quick purgatives. Cough should be allayed by opiates.

Astringents.—Such as:

Ergot, either subcutaneously, as ergotine, or by the mouth, in form of Extr. Ergot. Liq., ʒj, t. d.

Haseline, often of great value.
Gallic acid and opium.
Acetate of lead and opium.
Sulphuric acid and opium.

Alteratives.—Preparations of mercury or iodide of potassium when syphilis is present.

A blister over the affected portion of the lung is frequently of great use.

The after-treatment of hæmoptysis will depend, of course, upon its cause, but no patient should be allowed to return to his usual occupation until every trace of blood has disappeared from his sputum, and, if circumstances permit, entire rest, change of air, or a sea voyage should be advised.

III. Palpitation of the Heart

Palpitation is a symptom of the disturbance of the normal action of the heart. The term is sometimes applied to two varieties of disturbance,—the one, in which the heart beats rapidly but incompletely, and the other in which not only the frequency but also the force of the beats is increased. Properly speaking, however, the term should be restricted to the former condition. Increased force of the beats, with or without increased rapidity, is to be taken as a symptom of hypertrophy of the heart. It is seldom noticed or complained of by patients; is, indeed, more objective than subjective. Palpitation, as a subjective sensation, is a very common complaint, and must be looked upon invariably as an indication of a feeble rather than of an abnormally powerful heart.

Causes :—
1. **Weakness of the heart-muscle,** such as occurs in fatty and other degenerations of the heart. It is frequently associated with dilatation of the chambers. An atrophied heart also may be subject to palpitation.

2. **Mechanical obstruction to the cardiac contraction**, either occurring within the pericardium, as by an increase of the pericardial fluid, or without the pericardium, as by pleural effusion, consolidation of the lungs, or by distended stomach, enlarged liver, distension of the abdomen, or, again, by deformity of the chest or by tight lacing.

3. **Diminution of the blood pressure**, that is to say a diminished resistance for the heart to overcome. This occurs in anæmia, but in that condition there are other causes for palpitation, e. g. an insufficient or inefficient supply of blood to the cardiac muscle.

4. **The action of drugs** either upon the heart-muscle itself, upon the intrinsic ganglia, upon the vagus apparatus and centre, or upon the accelerator apparatus or centre. The exact way in which the various drugs act, as has been stated in the former chapter, is not quite known. We know, however, from experience that palpitation is frequently associated with tea-drinking and with tobacco-smoking in excess. The difference of the individual susceptibility to the action of these drugs is very marked: a single cup of tea will produce marked palpitation in one person, whereas half a dozen will not affect another.

5. **Nervous influence.**—From the excited action of the heart due to what is called **mere nervousness**, to the organic affection of the nervous mechanism produced by fright or shock. A variety of nervous palpitation is sometimes met with in persons who have been exposed to prolonged mental strain or worry, and is associated with insomnia and rapid wasting. The palpitation of Graves's disease should not be forgotten. Palpitation may be associated with hysteria or with epilepsy.

Symptoms.—Although palpitation is a symptom itself, yet it is a compound symptom, for with the rapid and excited action of the heart there are:

Subjective sensations of beating, fluttering, **rolling**, in the chest; a suffocating sensation in the throat; a tendency to gasp for breath or to **sigh**; great anxiety and a fear of impending death.

Cardiac pain—frequently relieved on pressure—sometimes approximating to angina pectoris.

Irregular or intermittent action of the heart.

Tendency to faintness with the usual signs of cerebral anæmia.

Palpitation may be a constant symptom, or it may come on in violent paroxysms after eating, or drinking tea, or smoking, or after some exertion or the like.

Treatment.—The treatment of palpitation must be directed to two **points**: (1) removal of the cause, and (2) alleviation of immediate distress.

    a. The most common cause of palpitation is, undoubtedly, **indigestion**. This is brought about by the ingestion of too much, too little, or **improper** food. Whether in this case the palpitation **is due** to the mechanical pressure upwards **of a** distended stomach, or to reflex **stimulation of** the cardiac ganglia, or to direct action **upon** the cardiac ganglia, it is our duty to carefully regulate the diet, to forbid the strong tea, **tobacco, or** whatever else is found by experience to produce dyspepsia, *e.g.* alcohol, uncooked fruit, nuts, and so on, and to administer purgatives, as constipation is so often present.

    b. In the other cases in which the palpitation is **not due** to organic disease, *e.g.* in anæmia and **debility, tonics**, especially iron (Ferri et Ammon. Cit. gr. viij) and strychnine (Liq. Strych. $\mathfrak{m}$v) **are of great** service, and some light alcoholic **stimulant in** moderate amount should also be **administered.**

    c. In **nervous palpitation**, due to mental strain and overwork, **too** much study should be forbidden, and exercise in the open air, change of air and scene, should be recommended. Nothing is better than a sea voyage to conduce to this end.

d. In **dilatation of the heart**, cardiac tonics should be given, especially **digitalis**, unless there be fatty heart or aortic disease, or other conditions to contraindicate its exhibition.

When it is not possible to remove the cause (2) **alleviation** should be attempted, and even when the cause may be **removed**, during the process of removal the following treatment should be tried.

Sometimes a **belladonna** plaster applied to the præcordia is all that is required. At others a warm linseed-meal poultice, with or without mustard, must be applied; but cardiac stimulants, such as ammonia, alcohol, **or ether are** often required, and the administration of opium (Tr. ♏v) in small doses may be tried with advantage. Certain drugs have a kind of special reputation as relieving palpitation associated with dyspepsia; they are, dilute hydrocyanic acid (♏iij-v), ether (♏x), with or without peppermint, lavender, cardamoms, or other **aromatics** and carminatives.

V. Angina Pectoris

[Suffocative breast pain; **epileptic neuralgia, asthma dolorificium**].

Angina pectoris was first described at length by Heberden in 1796, in a memoir 'De dolore pectoris' and since then has been the subject of much careful observation and study. It consists of a violent pain in the **chest** extending from the sternum to the spine, **nearly always** on **the left side, which comes on suddenly, and as** suddenly ceases, and is accompanied while it lasts with a feeling of approaching death. The pain may extend to one or both arms, most frequently to the left, and **may stop at the elbows.** More rarely it occurs in the **right arm and** sometimes in both arms. It may also reach to the fingers (Latham).

Conditions under which it arises.—From the conflicting accounts of the frequency with which angina pectoris occurs, it is evident that authors are not agreed as to what a typical anginal attack is. For example, Laennec states that angina pectoris in a slight or middling degree is extremely common, and exists very frequently in persons who have no organic affection of the heart or large vessels, whereas the majority of physicians hold that angina pectoris is very rare, and when it occurs is almost invariably associated with some definite organic lesion of the heart.

It is to the rare and severe or typical angina pectoris that we allude at present. The conditions with which it is chiefly associated are:

Valvular disease of the heart, especially incompetence of the aortic valves, whether produced by rupture, atheroma, or in any other way.

Fatty degeneration of the heart, particularly when the coronary arteries are diseased and atheromatous.

Aneurysm of the aorta.—Angina occasionally occurs with aneurysm of the aorta, but not very often.

Gout.—All the writers on the subject are agreed in connecting angina with gout and the high living and sedentary habits which give rise to it. Rheumatism is rarely a precursor of angina.

The predisposing conditions are:

Male sex.—It is very uncommon in women.

Late middle age.—Generally beyond fifty years.

High social position.—Possibly as likely to cause gout.

Symptoms.—The first attack as a rule comes on without any special warning of its onset, the exciting cause being frequently a very slight one, some **unwonted exertion** or some **sudden excitement** being sufficient to determine the paroxysm. **Walking uphill against the wind** after a meal is stated to bring on an attack; raising the arms to button a shirt collar, and any similar slight exertion may act in the same way.

It should not be forgotten that **the first attack** may be, and the **second or third attack**, if long intervals have elapsed between them, is frequently **fatal**.

If the patient survive the earlier attacks, it nearly always happens that for the future his attacks are of two distinct kinds, the one short, more frequent and less severe, and the other occurring less often, prolonged and severe.

When attacks are habitual, the patient generally has some warning of their approach. One patient of mine, whom I have seen several times whilst in an attack, nearly always said, "It's coming!" before the actual paroxysm took place.

The **attack itself**, which lasts from a few seconds to ten minutes or more, according to circumstances, comes on with increasing pain in the chest of most excruciating character, tearing, gnawing, or boring, accompanied by the characteristic sensation of suffocation and feeling of impending death. It is a **moot** point whether the patients are able to **breathe** during the attack. Some assert not only that they are able to do so but also that a deep breath relieves the pain, as pressure also does sometimes. In addition to the **directions in which the pain** is accustomed to radiate, as indicated in the above description of Latham's, it appears that many other nerves may be sympathetically involved, *e. g.* any of the branches of the brachial, lumbar, or sacral plexuses. The **pulse** during an **attack is almost** always affected; it may be full and **slow, or rapid, small,** and possibly irregular.

The **appearance** of the patient during an attack is almost characteristic. He has a look of intense suffering or horror; the eyes may be staring and fixed, the face pale and bathed with sweat. The termination of the attack is as sudden as the beginning; the patient takes a deep sighing inspiration, and relief is immediate. Sometimes **syncope** occurs during the paroxysm from which the patient is with great difficulty roused.

Varieties of pseudo-anginal attacks.—We cannot help thinking that it is best to restrict the term angina pectoris to the rare and severe form of paroxysm, as that is really the only one which presents all the characters indicated in Latham's description. The **other** forms are at least three in number, and may **be classed** under the generic name of pseudo-angina pectoris.

1. **Neuralgiæ**, especially cardialgia, pleurodynia, and gastralgia. In these affections, frequently connected with anæmia, dyspepsia, hysteria, or hypochondriasis, there is no evidence of implication of **the heart or** vessels; the attacks are irregular in course and duration. There is seldom any organic disease. Hope, a very able **writer** on angina pectoris, considers these attacks unworthy of the name of angina. In all probability many of Laennec's cases belonged to this class of neuralgias and were not true angina.

2. **Cardiac asthma.**—In this affection there is a peculiar interruption to breathing accompanied by pain of a **more or** less severe character in the præcordia. The breathing is laboured, rapid, shallow, and wheezy. The attack occurs suddenly and ceases suddenly, and is an occasional accompaniment of organic disease of the heart. The following case illustrates this variety of the affection:

A. Q—, aged forty-one, a thin, pale, and ill-nourished woman, was a patient of mine at the **Victoria Park Hospital** some time ago. She came to the hospital complaining of spasmodic attacks of difficulty of breathing *with pain*. Her chest was very deformed from lateral curvature of the spine, and the apex-beat was much displaced in consequence; a very loud systolic murmur was heard all over præcordia, chiefly at the base and **apex**; possibly, therefore, two different murmurs were present.

Although the dyspnœa or orthopnœa was spasmodic, the pain of which the patient complained was evidently not in her eyes anything like so important as the dyspnœa.

3. **Epileptic neuralgia** (Trousseau), or **syncope anginosa** (Laennec and Parry).—In this form there is a warning almost exactly like an epileptic **aura**, violent pain in the chest, and dyspnœa of most distressing character, and speedy unconsciousness, from which the patient may not recover. If he recover after the fit, stigmata or vibices, very similar to those observed after an epileptic fit, may be found in various parts of the surface of the body. Two cases of this kind are impressed upon my memory.

Two men suffering from aneurysm of the aorta experienced attacks which occurred suddenly, commencing with intense pain in the chest. The first, a sailor, aged forty, died in one of these paroxysms (the second during his stay in hospital). It was noted at the bedside that the attack was almost exactly like an epileptic fit. On autopsy it was found that the aneurysm had eroded several of the dorsal vertebræ, and that there was evidence of **pressure upon the left vagus nerve**, which was greatly flattened. In the second, **the** patient **also** died in one of these attacks, which was about the fifth during his stay in hospital; and here the **left recurrent nerve** was affected **as well as the left vagus.**

Attempts have been made **to** distinguish other varieties of attacks included under the term angina pectoris, but for all practical purposes the above description includes all the most important.

Pathology.—In considering the pathology it will

be best to confine our attention to the true form of paroxysm. The main theories of the cause of the anginal seizure are three in number:

1. **That it is due to spasm of the heart.**—This was the view that Heberden took, and in taking it he was followed by Latham. Heberden's reasons for believing in spasm of the heart as a cause were put forward at great length. Of these the most important may be cited as follows: That the attacks come and go suddenly; that there are long and complete intermissions between them, and that the pulse is unaffected during the height of the disorder.

Heberden did not, however, examine the bodies of more than one of the hundred cases he claimed to have seen. In this one "a very skilful anatomist could discover no fault in the heart, in the valves, in the arteries or neighbouring veins, excepting some small rudiments of ossification in the aorta."

We now know, however, that in cases of death from angina the coronary arteries are often diseased, and the muscular tissue of the heart is fatty. Under such conditions we are unable to see how the attack can be due to muscular spasm.

2. **That it is a pure neurosis or neuralgia.**—This is no doubt true of some of the varieties of angina as we have indicated, but this theory is inadequate to explain true angina. But under this head must be considered the important observations of Brunton. His researches are thought by some physicians to set at rest for ever the inquiry into the pathology of the condition.

Few observations had been previously made upon the pulse during an anginal attack. Heberden made the very astounding statement, as we have seen above, that the pulse was unaffected. Latham mentioned the pulse only casually. Brunton, however, observed the condition of the pulse in a case which, he says, "from the absence of a sense of impending death, might be reckoned as one of **pseudo**-angina, but in the intensity of the pain **and the manner of its radiation more closely resembled true** angina—as cardiac lesion was present, it belonged to the **class organic angina**"—and showed that in that **case** there was increased blood pressure. He supposed that this increase of the blood pressure was due **to a** spasmodic contraction of some, if not all, of **the** small systemic, and probably of the pulmonary arterioles, and he concluded that this was probably due to "a derangement of the vaso-motor system, and accompanied by a derangement of the cardiac regulating apparatus, producing **quickened instead** of slower pulsation."

In a case which I have recorded, I found the pulse, **when taken** during a paroxysm, amounted to **104 beats per** minute, and the tension was above normal. This patient suffered from aortic valve disease. **Other** observations have been made confirming Brunton's facts. He exhibited **nitrite of amyl,** a drug which diminishes arterial contraction and lowers the blood pressure, and demonstrated that it stopped the anginal pain. He showed, too, that other agents which diminish blood pressure, *e.g.* bloodletting, also diminish pain; and those which increase blood pressure, *e.g.* digitalis and aconite, **also increase** the pain. Since these most excellent **results** were obtained, this drug has been used **much** in cases of true and false angina with success. Without going into the apparent anomalies of the first recorded case, **as** Gairdner has done, which do **not, it seems to me,** interfere with the result, one **cannot** but agree with this writer when he says that "further observations seem to be required before it **can be** safely assumed that either the vaso-motor derangement on the one hand, or disorder **of the** cardiac innervation on the other, **is the primary or** essential phenomenon of true angina pectoris."

3. **That dilatation of the heart** is the cause of the paroxysmal pain. There are many reasons, we think,

for believing this is the correct explanation; of these three may be mentioned:

a. That the heart being generally in a condition of fatty degeneration is in the condition strongly predisposing to dilatation.
b. That the rapidity and not infrequently feebleness or irregularity of the pulse may be explained by supposing increasing dilatation of the heart. If the increased peripheral resistance precede the dilatation, it may be supposed that the weakened heart cannot overcome the increase of resistance to its stroke and so dilates.
c That the heart is almost invariably found after death in a condition of dilatation. This has been admitted by almost all, if not all observers.

Treatment.—We have now two drugs at any rate which seldom fail to relieve an attack of angina if administered quickly enough, viz. **Amyl nitrite** and **Nitro-glycerine** or **Tri-nitrine**.

The former is usually given to the patient to sniff, and for this purpose should always be kept in a well-stoppered bottle, and renewed frequently. Unless these precautions are attended to the very volatile substance loses its properties. Another way is to sprinkle a few drops on a handkerchief. In either case the patient should be directed to sniff **the amyl nitrite** until relief is obtained. This usually takes place as soon as the temples begin to beat and throb, and the face and forehead flush, and the head aches. These are indications that the capillaries and small arteries are dilated, the peripheral resistance diminished, and the arterial tension reduced. As a rule, a patient who is subject to anginal attacks, may be trusted to carry in his pocket a small bottle of the remedy, so as always to have it at hand if wanted.

The other remedy, **Nitro-glycerine** or **Tri-nitrine** introduced by Dr Murrell is also very efficacious, sometimes more so, at other times less so than the amyl nitrite. Its physiological action is apparently similar. This drug is now usually administered in the form of lozenges or tabloids, containing $\frac{1}{100}$ to $\frac{1}{50}$ gr. in each. The compound tri-nitrine tabloids containing tri-nitrine $\frac{1}{100}$ gr., nitrite of amyl $\frac{1}{4}$ gr., capsicum $\frac{1}{30}$ gr., menthol $\frac{1}{30}$ gr., appear to be a very convenient method of giving the two drugs together internally. The dose of the tri-nitrine may be gradually increased, but not too rapidly, lest nausea, pain in the stomach, and fainting ensue, which is sometimes said to happen.

In the cases which altogether resist the action of these drugs **opium** should be tried, especially in the form of subcutaneous injections of morphia. This remedy is especially valuable in cases not amounting to typical angina. We have used it largely in some cases with good success.

Local applications of mustard poultices, or of belladonna and opium liniments, are also recommended.

The treatment of the conditions giving rise to the angina would obviously be necessary after the attack was over; careful diet, **absence of** excitement, regulated and moderate exercise, purgatives, alkalies and the like, would be required on the one hand, change of air and scene, good diet, open air, tonics, especially iron and strychnine on the other, but in all cases careful attention to the digestion cannot be too strongly insisted upon.

VI. Syncope or Fainting

Syncope is a condition in which the blood pressure is suddenly diminished to such an extent that the brain and the central nervous system are insufficiently supplied with blood. This fall of blood-pressure is usually produced by a primary failure of the heart's action, although there are cases in which it appears that this is secondary to sudden diminution of the peripheral resistance.

Conditions under which it occurs.—1. Predisposing.—As a rule **young girls**, after puberty until twenty-three or twenty-four years of age, are most prone to **faintness** or attacks of syncope, especially if they are **anæmic** and **nervous**.

2. **Anatomical.**—If what has been said in the first chapter as to the essentials of a natural cardiac action has been followed, the conditions under which syncope may arise will be easily understood. Of these the most important are as follows:

   a. **Insufficient supply of blood within the chambers of the heart** especially occurring from—
      1. Severe and sudden hæmorrhage to such an amount as to appreciably diminish the total volume of blood in the body, *e. g.* post-partum hæmorrhage, rupture of a large vessel or of an aneurysm; in many such cases syncope simply precedes death.
      2. Retention of too large an amount of blood within the venous system. It has **been** called bleeding into the veins, **and may** occur when pressure upon the outside of **veins** is too suddenly **and completely** removed and the inside pressure is very small. This is particularly likely to take place when the abdominal veins are thus released, **as when ascites** is relieved by tapping and pressure is not kept up with a binder. It cannot be too strongly insisted upon that the capacity of the abdominal veins is equal to the total amount of blood in the body. **Fainting** may also happen in paracentesis thoracis.

      Obstruction of the return of blood to the heart, in the lungs, or elsewhere, may have a like effect.

b. **Malnutrition of the heart's substance**, and this must be taken to include not only the muscle but also the ganglia. Such a consequence results from—
   1. An insufficient supply of duly arterialised blood, as in anæmia, phthisis, and other wasting diseases.
   2. The circulation of poisoned blood, acting either upon the muscles, the nervous apparatus, or upon both.
      The most common effect of both these circumstances is fatty degeneration, the frequent precursor of dilatation.
c. **Inhibition of the heart action** through the central nervous system or a sudden either local or general dilatation of the arterioles. When these nervous phenomena are witnessed, they are nearly always instances of reflex inhibition. The sensory nerve terminations stimulated being often those of special sense, a bad smell, a horrible sight, &c., sometimes causing syncope. It is difficult to separate such cases from cases of shock.
d. **Diminution of the arterial tension** from any cause including (c), and so a diminution of the resistance the heart has to overcome. This, it must be recollected, may be either the cause or the effect of the inefficient cardiac stroke. Thus, it may be due to (i) a dilatation of the arterioles, e.g. by heat, as in a warm bath, by drugs, or by central vaso-motor inhibition, as above mentioned; (ii) by rapid bleeding; or (iii) venous dilatation; or (iv) regurgitation of blood through the aortic valves during diastole.
e. **Mechanical obstruction to the action of the heart**, such as occurs in pericarditis from effusion or adhesions, or from pressure of a distended stomach, an enlarged abdomen, and so on.

f. **Dilatation of the heart,** either sudden or gradual, and produced in any way, **must be** considered as exceedingly prone to be associated with syncope.

Thus we see **that the** cause or result of all these conditions **is a** temporary fall of blood pressure, **together with** partial anæmia of the brain, and unconsciousness.

These in general produce syncope; they are probably instances of reflex stimulation of the cardio-inhibitory or the vaso-motor centres.

Phenomena of syncope.—There are as a rule **premonitory symptoms** of a fainting fit, **languor, restlessness, muscæ volitantes, noises in the ears, coldness and shivering, sighing, pallor of the face, perspiration on the forehead, a feeling of nausea, trembling of the legs.** These may quickly give place **to complete** unconsciousness, with dilated pupils, in **which state** the patient may **without** warning, fall down unconscious, the surface of the body and face being bloodless, and frequently covered with sweat; pulse scarcely to be felt, but if so, small and very rapid; respirations also appear to **be slowed.** Sometimes associated with these symptoms are more or less well-marked twitchings of the muscles. As a rule, the heart-sounds are **just audible.**

The **severity** of **the** syncope varies from the exceedingly frequent form to which anæmic females are subject, often brought on by the hot and stuffy atmosphere of a crowded church or concert room, and from which they usually quickly "recover" on **removal** into the **open** air, to the severe form from **which** patients **are** with difficulty roused, and which, indeed, **is** frequently fatal, arising, it may be, from some organic heart disease. The duration, too, may vary from a few seconds to half an hour or more (Dr **Hope** says *days*).

When recovery takes place, the patients **are in a** dazed **condition for** a longer **or** shorter time, the colour gradually returns to their faces and lips, and they are able to stand with some assistance, but they have no recollection of anything which has occurred during their attack.

Treatment.—A change from the standing to the sitting position will often ward off a fainting attack, especially if the patient sit with his or her head inclined below the level of the body. Removal into pure air, and the application of smelling salts or similar substance to the nostrils, or drinking cold water, will often recover a person from commencing syncope; or again, the administration of stimulants such as alcohol or ether. Sprinkling the face with cold water is also to be recommended. In cases of more marked syncope the application of a hot poultice with mustard, over the cardiac region is sometimes attended with benefit, or other stimulant remedies applied locally; pressure upon the abdominal aorta or other large arteries may be tried.

When syncope arises from severe hæmorrhage it is possible that nothing but transfusion will produce permanent benefit; in that condition often one faint follows quickly upon another. When syncope arises from organic heart disease or from chloroform or other poison, galvanism over the cardiac region should be attempted. The subcutaneous injection of atropine has been used with success.

As, however, the most common condition predisposing is anæmia and debility, the ordinary treatment of these affections by iron and other tonics, good diet, with or without alcohol, change of air, and regulation of the bowels, is very probably necessary.

CHAPTER III

THE PHYSICAL EXAMINATION OF THE CHEST

In every case in which disease of the thoracic viscera may be reasonably suspected, it is necessary to have recourse to a physical examination of the chest. This examination comprises four methods, named respectively:—1. **Inspection**; 2. **Palpation**; 3. **Percussion**; 4. **Auscultation**; and these should be used **in order** in every case. In this chapter the application of the several methods will be considered as briefly as possible.

I. **Inspection** is the method of examination by the eye alone, and consists of looking at the chest with a view of detecting any evident alterations in its appearance. Attention should be specially directed in the examination to any abnormalities:

A. **Of the skin and subcutaneous tissue covering the chest**

Normal condition	Abnormal conditions
Should be supple, elastic, of the usual complexion of the individual in health, and should be free from all eruptions.	The skin may be abnormally: **White** in anæmia from any cause, and also sometimes in wasting diseases, e. g. phthisis. **Yellow** in jaundice or pneumonia. **Pigmented**, either generally or locally, e. g.

about nipple, or about the site of old blisters; this may occur in Addison's disease, but also in chronic wasting diseases.

It may be covered, more or less, with skin eruptions, specific or nonspecific, but especially with **sudamina** or **miliaria** in acute rheumatism and other fevers—or with **Tinea versicolor**, a parasitic disease (parasite = *Microsporon furfur*), chiefly down the sternum, over the mammary regions, and on and between the shoulders, the patches of the eruption being of various sizes, from a line to several inches in diameter; they are slightly raised, of sinuous outline, and often covered with minute branny scales. This eruption is common enough in health, but is very frequent in cases of phthisis, &c.

The **veins** should not be evident, **except in** pregnant and nursing women, in whom those passing from the breasts are large, blue, dilated, and tortuous.

The **veins** are sometimes very evident, dilated, and tortuous, chiefly in **cases of** obstruction to the intrathoracic veins by aneurysms or morbid growths. The **capillaries** and **venules** may be dilated in patches or in lines, in cases of obstruction to the pulmonary circulation, as in emphysema.

The subcutaneous tissue may be evidently **œdematous** in cases of local or general venous obstruction, or in general dropsy from kidney or liver disease. Subcutaneous **emphysema** may also be evident, but it is most likely to be detected by palpation.

B. Of the Muscular System and General Nutrition of the Chest

There should be sufficient subcutaneous deposit of fat to cover the bony prominences. The muscles should be firm and well developed, but the exact normal condition differs of course in the individual.

There may be **too much** fat covering the chest, the muscles are then as a rule wanting in tone; this is apt to arise in persons of sedentary habits, and in those who indulge freely in alcohol, and in women past the child-bearing age.

Wasting of the fat and muscles is apt to occur in phthisis, carcinoma, and diabetes mellitus.

C. Of the Shape and Size of the Chest

Differences in individuals even in good health and strength. A typical chest should be

Abnormalities consist in—
1. Dilatations.
2. Contractions.

symmetrical, both sides being equal in circumference and alike in form. It should arch forwards slightly, the arching increasing from below the clavicles to the fifth rib, and then receding gradually to the lower ribs. There should be no supra- or infra-clavicular depressions, neither should the upper ribs be evident. The angle of Ludwig, or that made by the upper bone with the body of the **sternum**, should not be exaggerated. The nipple should be on the level of the fifth rib or fourth intercostal space. The pectoral fold of skin, corresponding with the lower border of the pectoralis major, should be evident in a muscular subject.

1. **Abnormal dilatations**:
   a. **Unilateral**, occurring in:
      i. **Pleural effusion**, the pleura being distended with fluid, and bulging, chiefly at the base.
      ii. **Pneumothorax**, the pleura being distended with air.
      iii. **Hypertrophy of the Heart**, or **Pericardial effusion**, may produce local bulging of præcordia.
      iv. **Aneurysms**, Tumours of the bones, *e. g.* of the sternum, of the lungs, **liver**, or spleen, may cause local distension.
   b. **Bilateral** exists in the so-called "barrel-shaped" chest of emphysema, in which the diameters of the chest in general are increased. Sometimes the upper part of the chest is chiefly affected.
2. **Contractions** of the whole or part of one side occur chiefly in chronic pleurisy, after the absorption or evacuation of the fluid effusion. Contractions, depressions, or flattening of the chest, of less extent, occur in the various forms of wasting diseases of the lung, especially in fibroid phthisis, the commonest

seat of contraction being above and below the clavicles.

In addition to the above these are the following types of what may be called **subnormal chests.***

1. **Alar or pterygoid.**—Narrow and shallow, antero-posterior diameter reduced, length increased. The ribs are more than ordinarily oblique, shoulders full, and the scapular angles project like wings.
2. **Flat.**—A condition somewhat similar to 1, but the shape is changed by the true ribs losing their curve and becoming straight. The front of the chest looks flat, or the sternum may be even depressed backwards so that the costal cartilages curve in the wrong direction.

   1 and 2 indicate a tendency towards phthisis.
3. **Transverse constriction of the chest.**—A more or less well-marked depression of the lower portion of the chest wall in front, from the xiphoid cartilage to mid-axillary line. This is common. It is due to some impediment to the free entrance of air into the lower portion of the lungs. Its cause usually is catarrh of the lungs in infancy.
4. **Pigeon breast.**—This is due to a straightening of the true ribs in front of their angles, the sternum is thrown forward, and the greatest lateral diameter of the chest is thrown backwards. The horizontal section thus tends towards a triangular form. With this is nearly always associated a considerable amount of lateral constriction.
5. **Rickety chest.**—In this variety, a shallow

* After Gee.

longitudinal groove is found on either side of the chest, at the junction of the ribs and the costal cartilages.

D. Of the Respiratory Movements

Normal
Regular in time and in depth, equal on both sides; of the inferior costal or costo-abdominal type in man (the intercostal muscles being little used), and of the superior costal type in women and children.

Abnormal
The breathing may be unequal on both sides—

1. When this is the case it is noticed that one side partially or wholly moves less freely than the opposite side.

When the upper part moves badly we usually have to do with some form of wasting disease of the lung.

When the lower part moves but little, a pleural effusion or pneumonia may be suspected.

If the whole side be motionless pneumothorax is likely, or extensive pleural effusion, or very advanced phthisis.

b. Depressions of portions of the chest wall occurring in inspiration, with bulgings during expiration—the reverse of normal—occur in the upper intercostal space in phthisis, especially when a cavity is present, in the middle spaces in adherent pericardium, and in the lower part of the chest and epigastrium and elsewhere in severe emphysema, but especially in croup and other laryngeal obstructions in children.

The type of respiration may be changed, as when the diaphragm is unable to act from pressure of the abdominal contents or from paralysis.

Rapidity.—14 to 16 respirations per minute.

The number of respirations may be increased to 30, 40, 60, or even higher. This increase occurs when from any cause dyspnœa is present (see p. 52).

A rhythmical irregularity of breathing, known as **Cheyne-Stokes' respiration**, is observed in the later stages of some diseases of the heart, especially in fatty degeneration, as well as in some other diseases. It consists of a regular series of rapid inspirations, increasing regularly in depth until an acme is reached, when the inspirations become gradually slower and shallower until they cease altogether, and a pause of about the same duration as the series of inspirations ensues, when a similar cycle of respirations begins again.

E. **Of the Apex-beat**

Normal

Position.—Two inches below the left nipple and one to the right of it. In the fifth interspace, between the left parasternal

Abnormal

The apex-beat may be displaced either to the left or to the right.

Displacements to the left occur chiefly in—

and mammary lines. It may be in the fourth space in children.

i. Hypertrophy of the ventricles.
ii. **Air or fluid in the right pleura.**
iii. Tumours to the right of the heart.
iv. Shrinking or contraction of the **left** lung, especially in chronic pleurisy.
v. **Pressure upwards** of the diaphragm from abdominal affections.

The amount of displacement varies; generally it is downwards as well as outwards, but in iv it may be upwards and outwards.

Displacements to the right occur in conditions similar to the above (ii, iii, **and iv**). The apex may beat to the right of sternum also, in the rare event of the viscera of the chest being transposed.

Character and extent. —Distinct, yet not heaving. Localised to a small space **not more** than an inch.

Cannot be seen, as in very stout people, in œdema of the skin, or in emphysema.

May be heaving, forcible, and extensive, as in hypertrophy of the heart.

May be diffused without being forcible, in pericarditis.

Look also for any abnormal pulsation.

Abnormal pulsation is observed—

  i. In cases of **aneurysm of the aorta or** innominate artery, generally in the neighbourhood of the second right interspace.

  ii. Pulsation to a less marked extent may occur in **aortic disease**, without aneurysm.

  iii. A **dilated right auricle** may produce a systolic impulse in the fifth interspace to right of sternum, and also in the epigastrium.

II. **Palpation** is the method of examination by the hand. It is used **to confirm** the observations made by inspection. Any abnormalities detected by the latter method should be further investigated therefore by palpation. There is, however, special information to be obtained by palpation with respect to the vibrations of the chest wall under certain conditions.

A. **The vocal fremitus.**—The chest wall is thrown into vibration during speech, and its vibrations may be felt if the hand be applied to the chest. **The cause of the fremitus is that the vibration of the vocal cords during speech is communicated by the column of air in the trachea and bronchi to the pulmonary air-vesicles, and so to the chest wall.**

The strength of the vocal vibrations depends upon **the pitch of the note**; the deeper **the** note the more marked the vibration, so that with bass voices and in men it is most marked. It diminishes as the **distance from the larynx increases**; and is felt **more on the right side** than the left, because the right bronchus is wider than the left, and is given off at a less acute angle. Excessive development **of the** chest coverings, its fat and muscles, acts as a damper and diminishes the vocal fremitus. In pathological states the vocal fremitus may be (*a*) **increased** or (*b*) **diminished**.

(a) Conditions in which the vocal fremitus is **increased**:
  i. When the **lung is consolidated** from any cause, but especially in pneumonia, the vibrations of the column of air in the nearest bronchus are conducted excellently by means of the consolidated lung to the chest wall.
  ii. Over **cavities** with dense walls near the surface, freely communicating with a bronchus, the vocal vibrations of the air in the bronchus are even amplified.
  iii. In **emaciated individuals.**
(b) Conditions in which the vocal fremitus is **diminished** or **absent**:
  i. In **pleural effusions of any kind**, if the effusion be slight there is diminished fremitus, and if large the fremitus is absent over the position of the fluid. The fluid acts as a damper, and stops the conduction of the fremitus to the surface.
  ii. In **pneumothorax**, the air acts in the same manner as the fluid in i.
  iii. In **obstruction to a bronchus**, diminishing its calibre partially or wholly, the fremitus may be diminished. iv. In **stout** people.

B. **Friction fremitus.**—The grating of one roughened surface of the pleura over the other in pleurisy may sometimes be perceived during inspiration on the application of the hand.

C. **Rhonchial fremitus** is sometimes present in bronchitis when there is considerable obstruction from swelling of the mucous membrane or from accumulation of tenacious secretion. The air in the tubes is thrown into extra vibration by the vibration of the bronchial walls and of the fluid contained within them.

D. **Cavernous fremitus.**—In cases of vomica in the upper lobe of the lung, if sufficiently superficial, a special fremitus due to the vibration of the fluid in the cavity may sometimes be felt.

E. **Fluctuation.**—In cases of excessive pleural effusion fluctuation of the fluid may occasionally be detected if one hand be placed on the side, and the surface of the chest be tapped with the finger in an intercostal space at some little distance.

Palpation of the cardiac region.—In the cardiac region palpation should be directed to further investigating the position and character of the apex-beat, and of the abnormal pulsations noticed on inspection. It frequently happens that the position of the **apex-beat** and abnormal pulsations not made out by inspection alone may be detected by palpation.

In addition to those systolic impulses noticed on inspection, with palpation **præsystolic** and **diastolic** impulses may occasionally be detected, the former in cases of marked hypertrophy of the left auricle in the fourth interspace near to the sternum, and the latter either at or near the apex-beat in hypertrophy of the heart or in the left second intercostal space near to the sternum; when in the latter situation it usually corresponds with an accentuated second sound.

Impulses in the cardiac region may be accompanied by **thrills**, a sensation being felt when the hand is applied to them such as would be perceived upon stroking the back of a purring cat; hence these thrills have been called **purring tremors** or **frémissement cataire.**

Thrills or **purring tremors** felt in the following situations, if well marked, indicate definite conditions and correspond with **murmurs** heard loudest in the same situations.

I. **At apex-beat** .
  1. Systolic thrill = Mitral incompetence.
  2. Præsystolic or diastolic thrill = Mitral stenosis.

II. At the second right intercostal space, close to sternum . . .
  3. Systolic thrill = Aortic stenosis or obstruction.
  4. Diastolic thrill = Aortic incompetence.

III. At the ensiform cartilage, or over fourth costal cartilage or interspace, close to the sternum . . .	5. Systolic thrill = Tricuspid incompetence. 6. Præsystolic or diastolic thrill = Tricuspid obstruction or stenosis.
IV. At the second left interspace, close to the sternum or slightly below	7. Systolic thrill = Pulmonary obstruction or stenosis. 8. Diastolic thrill = Pulmonary incompetence.

Of the above eight thrills Nos. 2 and 3 are by far the most common and diagnostic. Systolic mitral thrill I consider to be very uncommon.

Pericardial thrill or friction fremitus.—In some cases of pericarditis the rubbing of one roughened surface of the pericardium against the other produces palpable fremitus.

Fluctuation of large pericardial effusions may be sometimes detected.

Thrills may be sometimes felt over pulsating aneurysmal tumours.

III. **Percussion** is the method of examination by striking or tapping the chest so as to produce sounds, from the characters of which the condition of the subjacent organs may be inferred. It is either **immediate**, when the chest is struck directly with the ends of the fingers, or with the flat of the hand, or **mediate** when something intervenes between the chest wall and the percussing finger or fingers.

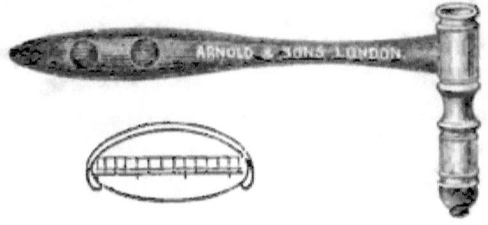

FIG. 12.—Plessor and Pleximeter.

This something is generally the first finger of the opposite hand, but a thin plate of ivory may be used (**pleximeter**). A hammer or **plessor** too may be substituted for the fingers in percussing, but without much gain.

As regards the number of fingers used, practice differs; some use three, others two, and others one, either the first, second, or third finger. The movements of the hand in percussing should be from the wrist with the arm motionless.

The sound produced by the percussion of the chest posseses two of the properties of a musical tone, viz. **intensity and pitch.** Pitch depending upon the number of vibrations in a given unit of time, and intensity upon the extent of the vibrations. A musical tone, it should be remembered, is made up of rapid rhythmical or isochronous vibrations. If the vibrations are not rhythmical the sound is not a tone but a noise. In these sounds, therefore, we have to deal with **intensity, pitch,** and **tone.**

1. **Intensity** of percussion note, *i. e.* the loudness; the terms indicating this property are loud and soft. It is not a very important part of the sound produced, as it depends chiefly upon the force exerted in percussing.
2. **Pitch.**—According to their pitch, percussion sounds are named in order as the number of vibrations increase.
    - α. Low-pitched or **tympanitic,** the sound produced in percussion over the distended belly.
    - β. **Subtympanitic** over normal lung.
    - γ. **Tracheal,** as over the trachea.
    - δ. **Osteal,** the sound yielded by the percussion of bone, cartilage, and solid structures.
3. **Tone.**—The percussion sounds are seldom pure musical tones. Their musical properties may be obscured, muffled, or dulled. The terms used to indicate their musical property or its diminution or absence are **clear, muffled,** or **dull.** A

dull sound of the percussion note is one without tone or a noise.

(**Fulness**, and its opposite, **scantiness**, are said by Gee to indicate the duration of the note, which in turn depends upon the elasticity of the percussed body.)

A very considerable amount of confusion exists as to the exact meaning of the terms used in describing the percussion note.

The sub-musical tone may possess high or low pitch, loudness or softness, fulness or scantiness.

A dull sound or a noise does not possess pitch, but may be loud and full or the reverse.

Cause of the percussion sound.—How the percussion sounds of the chest are produced is a disputed point. There are several theories.
1. That it is due wholly to the vibration of the chest wall.
2. That it is due to the vibrations of the air within the pulmonary alveoli.
3. That it is due partially to the vibration of the chest wall, and partially to the air within the pulmonary alveoli, modified by the state of tension of the lung tissue.
4. That the lesser bronchial tubes yield the percussion sound, which is conducted by the tissues solely, the qualities of the tone depending upon the amount of their thickness and elasticity. When the conducting material is thick and inelastic, the sound is more muffled, lower-pitched, and less loud than when the reverse is the case.

This last theory is the one adopted by Gee and is certainly the most workable.

The Normal Percussion of the Chest.—The percussion sound is subtympanitic over the whole of the right chest in front to the sixth rib, being a little less resonant over the apex, and in the fifth intercostal space than elsewhere. Below the fifth interspace the sound is quite dull from the situation of the liver upon that side. Upon the left the sound is as on the right, except that there is a small, more or less triangular patch of dulness, corresponding to the heart lying subjacent uncovered by the lung. This area of cardiac dulness begins above at the junction of the fourth rib with the sternum, and is bounded on the left by a line drawn from this point to the apex-beat, and on the right by a line drawn down the left edge of the sternum. Below, the dulness may be considered to be included by a line from the apex-beat to the sternum at right angles, but in reality it is difficult to say where cardiac dulness ends and liver dulness begins, as they are continuous in this situation. Over the sternum also the sound given by percussion is subtympanitic. At the level of the sixth rib on the right side the tympanitic note of the stomach first appears.

Laterally, the sound is subtympanitic on the right side to the eighth rib, and on the left to the ninth. Behind, the sound down to the tenth rib on either side is subtympanitic, but not quite so clear as in the front of the chest. In the supra- and infra-spinous fossæ, and close to the spines of the vertebræ, it is more high pitched (? osteal) than elsewhere.

A. Conditions producing increase in the resonance on percussion

1. **In pneumothorax** the note being tympanitic but not clear.
2. **In cavities** almost if not quite empty of fluid and near the surface, or deeper if with firm tissue between them and the surface, provided they are of fair size and in front of a large bronchus.
3. **In conditions when the lung tissue is relaxed**, e.g. (a) in pleural effusion the dulness of extensive effusion, preceded by a clear percussion note, though high-pitched; and, moreover, in moderate effusion a similar percussion note is

observed above the limit of the effusion in the front and side of the chest. (*b*) In the first stage of pneumonia, and when it is undergoing resolution, in œdema, congestion, tubercle, &c., where the air-vesicles are relaxed from the presence of fluid or solid material within them. In all these cases the note is high-pitched.

4. **In emphysema** not only is there diminution of the cardiac and liver dulnesses, but also the percussion note is over-resonant though not clear.

B. Conditions producing decrease in the resonance on percussion

1. **Consolidation** of the lung from any cause without relaxation of its tissue, *e.g.* pneumonia, phthisis, infarct, &c.
2. **Pleural effusions,** if the effusion be sufficient. It should be remembered that there is not absolute dulness in the **early stages.**
3. **Collapse of the lung.**
4. **Extreme stretching of the air-vesicles** as in violent straining inspiration.
5. **Cavities** in the lung, which are surrounded with tough, unyielding walls, or which contain much fluid.

Special percussion sounds

1. The **cracked-pot sound or bruit de pot fêlé.** This is a more or less metallic sound, which may be imitated by joining the hands with the palms inwards so as to include a hollow space, and by knocking the back of one of them against the knee a sound like that which arises from the clinking of money is produced. The conditions necessary for the production of this sound seem to be the sudden escape of air from an enclosed space under pressure.

The cracked-pot sound may be produced on percussion of the chests of children when screaming, of adults when singing a prolonged note, or if the chest be covered with much hair.

In pathological conditions of the lungs the sound is produced on percussion:

- *a.* Of **cavities of medium size**, freely communicating with a bronchus, on firm percussion, the patient's mouth being open.
- *b.* Of **pleural effusion**, at the upper limit of the fluid.
- *c.* Of **pneumonia**, in the neighbourhood of consolidated parts.
- *d.* Of **pneumothorax**, with opening in the thoracic wall.

2. **Amphoric or metallic resonance**, is like the sound produced on percussing an empty jar or bottle.

The sound arises on percussing over—

1. **Cavities in the lung substance**, which must not be smaller than 6 cm., and must have smooth, curved walls.
2. **Pneumothorax**, when the tension of the gas in the pleura is of a certain amount.

The area of **Cardiac dulness** is diminished in emphysema, and increased when there is retraction of the left lung in phthisis and old pleurisy, as well as in pericardial effusion and enlargement of the heart. In this latter condition, as well as in some other affections, the dulness may be displaced as well as increased.

Percussion resistance.—By this term is meant the sensation conveyed to the finger of the percussor, of yielding or not yielding of the part percussed. The resistance depends upon the compressibility of the part, and hence is greater when the fluids or solids are percussed than when the part percussed contains air.

IV. **Auscultation** is the method of examination by listening to the sounds within the chest, either by the direct application of the ear or through the medium of a special instrument called a stethoscope. The former method is called **immediate** and the latter **mediate** auscultation. Immediate auscultation is little practised, except under special circumstances, at the present time. It is inconvenient, and possesses few if any advantages over the mediate method.

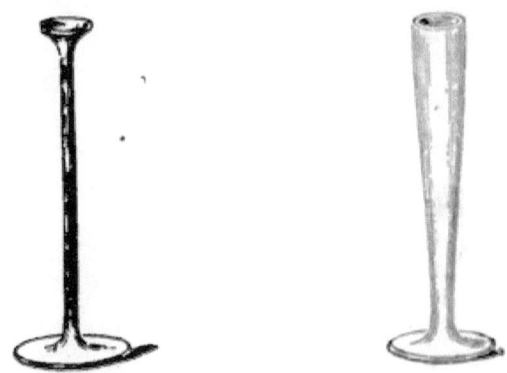

FIG. 13.—Examples of Wooden Stethoscopes.

Stethoscopes are of various forms. The simplest (Fig. 13) consist of a solid rod of wood with an expanded end or ear piece. Generally speaking, however, the rod is not solid but tubular. The end of the instrument applied to the chest is often a small trumpet-shaped expansion, about an inch in diameter; the tube itself is about one third of an inch in diameter, and about seven or eight inches long; the ear piece, about two inches or more in diameter, too, is often slightly hollowed out to receive the ear most conveniently. Instead of wood other substances, such as metal or vulcanite, may be substituted.

Stethoscopes, such as are above described, are for use with one ear only. Binaural stethoscopes are now much employed. Their usual structure consists of two metal or vulcanite tubes about two inches to one foot long, curved at one end nearly at righs angles, the curved ends fitted with ivory ear-piecet made as just to fit the auditory meatus; the other extremities are fixed in a metal frame, so that one arm can be separated from the other. To the fixed ends are attached flexible tubes of thick india rubber about eight to ten inches long, which in turn are attached to a metal, caoutchouc, or wooden chest piece three or four inches long.

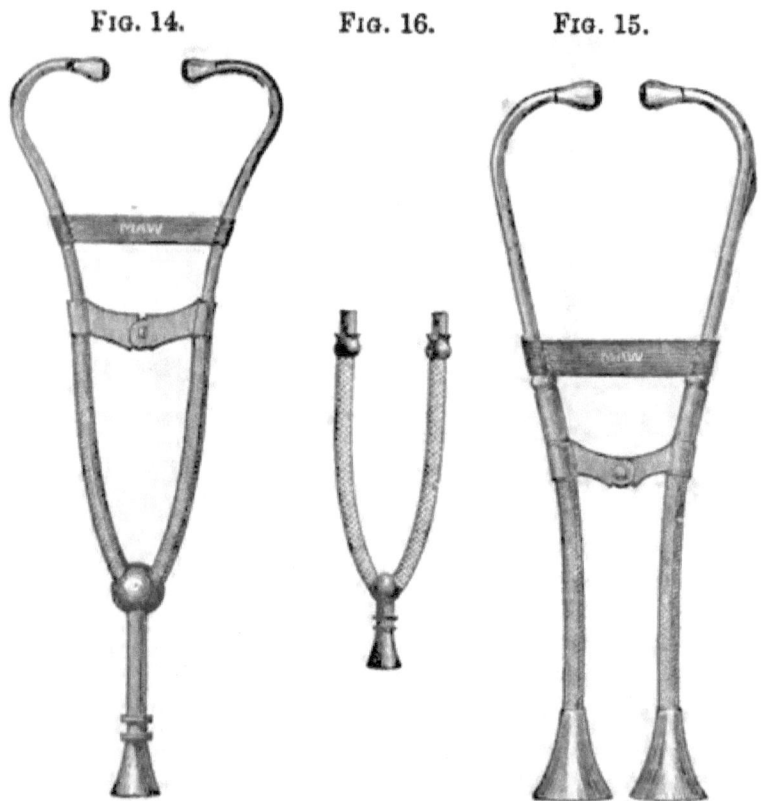

FIG. 14.—Binaural Stethoscope.
FIG. 15.—Differential Binaural Stethoscope, with removable ear-piece. It may be altered into the ordinary form by removing the two end-pieces and substituting one (Fig. 16).

The choice of a stethoscope is a matter of individual opinion. To my mind the binaural stethoscope is the most useful in the investigation of chest diseases, although no doubt practice with it is necessary in order that its tendency to exaggerate and modify sounds may be discounted.

Auscultation of the chest may be divided into four parts.

1. Of the breath-sounds.
2. Of the voice.
3. Of the pleural sounds.
4. Of the heart and great vessels.

1. Auscultation of the breath-sounds. — If the stethoscope be applied to the chest in health, a faint sound or rustle, or faint sighing murmur, is heard corresponding in time to the inspiratory movement, but also extending over a part of the expiratory movement; this sound is known as the **vesicular murmur**. It is heard louder when the respirations are increased in number and depth, and its expiratory portion too is then plainer. When the vesicular murmur is louder than normal it is called **puerile**. Frequently the only part of the murmur which is audible is the inspiratory.

The inspiratory sound may be **interrupted**, or jerking, or **deferred**, and the expiratory sound may be prolonged, when expiration is prolonged, without much significance.

Bronchial breathing is the sound which is heard on listening directly over the larynx, trachea, or bronchi. It can be distinguished on auscultating between the scapulæ, particularly near the upper and inner angles, or from the seventh cervical to the fourth or sixth dorsal vertebra behind, in many perfectly healthy people, if not too stout. In this variety of breathing, the expiratory sound is almost if not quite as audible as the inspiratory, but both are prolonged. The special character of the breathing can scarcely be put into words. Both inspiratory and expiratory sounds are said to resemble the utterance of the aspirate H, or the sound produced by blowing down a tube of the same bore as the main bronchus. Gee calls it **hollow** or **reverberating**. To Laennec it suggested a drier **sound** than the vesicular murmur, and the idea that air is passing through large empty space.

As modifications of bronchial breathing, it is necessary to mention **tubular breathing**, a sound as of the transmission of air in a tube rather smaller than a main bronchus, heard typically over hepatised lung, and **cavernous breathing** which is an exaggeration of bronchial breathing, having the same characters, but conveying an idea of air entering into a larger space than a bronchial tube.

Blowing or puffing respiration is described by Laennec as occurring with bronchial or cavernous breathing when the patient breathes rapidly ; the air during inspiration appears to be drawn from the auscultator's ear, and during expiration to be blown into it. The sound is heard both in superficial and in the deeper bronchi in pneumonic consolidation or pleural effusion, also in superficial **vomicæ**.

The vesicular murmur.—In health the vesicular murmur is heard all over the chest, excepting at that part where the respiration is bronchial, viz. between the scapulæ behind. It is heard more distinctly in front than behind, and on the right side rather more distinctly than on the left. Below the clavicles it is plainest, and over the scapulæ most obscured.

Cause.—Opinions differ. The chief suggestions are:
1. That it is due to the entrance of air into the alveoli during their dilatation.
2. That it is due to the vibration of the lung substance from its increased tension during inspiration.
3. That it is the sound produced in the glottis modified by conduction through the spongy lung substance.

In disease the vesicular murmur may be **weak**, or

may be replaced by bronchial or other form of respiration. The weakness is usually due to insufficient breathing, but may be due to inefficient conduction.

Bronchial breathing is the laryngeal sound as heard in the bronchi conducted to the surface without alteration or diminution. This kind of respiration, or one of its above-mentioned modifications, occurs over:

1. Any kind of consolidation of the vesicular substance, *e. g.* from inflammation.
2. Cavities, *e. g.* phthisical or bronchiectatic.
3. Collapse of the lung, *e. g.* in pleural effusion.

Sounds added to the respiratory murmur— râles

These are sounds occurring with the breath-sounds only in morbid conditions. They are of two kinds, (*a*) dry and (*b*) moist sounds.

(*a*) **Dry sounds.**

1. **Rhonchus and sibilus..**—These sounds are due to the diminution of the calibre of the air-tubes, rhonchus in the large, and sibilus in the small tubes, the former either low pitched or sonorous, *e. g.* snoring, cooing, groaning, and the latter squeaking, whistling, and the like, high pitched or sibilant. The usual cause is bronchial catarrh, wherein the mucous membrane is swollen or is covered with mucus.

2. **Creaking or crackling**, or the smallest variety of crepitation, called also the **crepitant râle**, is a highly characteristic inspiratory sound occurring in the early stage of acute croupous pneumonia and also in œdema of the lung. It can be closely imitated by rubbing some hairs between the finger and thumb close to the ear. It is also likened to the crackling of salt when thrown on the fire. The theory which supposes that this sound is due to the opening out of collapsed air-vesicles is probably the correct one.

3. **Dry crepitation**, is a sound occasionally heard in emphysema, due most likely to a cause similar

to (2), *i. e.* the stretching out of emphysematous air-vesicles.

(*b*) **Moist sounds.**

Crepitations or mucous râles, the sounds made by the passage of air through fluids within the lung. The varieties depend upon the size of the air-bubbles. Thus there are crepitations of all sizes, from the large crepitation to the small crepitation; generally speaking, the former are produced in the large tubes, and the latter in the small tubes. The variety of crepitation occurring in a cavity is sometimes called **gurgling**, and the smallest kind, when akin to pneumonic crackling, is called the **subcrepitant râle.**

(2) **Auscultation of the voice-sounds.**—On auscultating the healthy chest during phonation, nowhere except in the situations where bronchial breathing is heard can the voice be distinguished except as a mere buzzing. The words spoken cannot be recognised.

Under conditions which are for all practical purposes the same as in the case of bronchial breathing, the voice, as heard in the trachea or bronchi, is conducted to the surface of the chest without alteration or diminution, and **bronchophony** is heard. It is, *cæteris paribus*, loudest in the interscapular space behind than elsewhere.

The physical states which produce bronchophony may be summarised as follows:

1. Consolidations of the lungs of whatever kind.
2. Vomicæ of the lungs with firm walls.
3. Compression of the lung by fluid. Bronchophony is heard over the situation of the compressed lung.

Pectoriloquy is an exaggerated or intense bronchophony which is heard over large and superficial vomicæ in the lungs. It differs only in degree from ordinary bronchophony. No doubt, however, that the voice as heard over a very large vomica has sometimes a peculiar metallic or amphoric character, but this is not necessary to pectoriloquy.

Ægophony is a peculiar character of the voice which is likened to the bleating of a goat or the voice of Punchinello. It was considered to be diagnostic of pleurisy with moderate effusion. It may, however, be found under other circumstances, although less commonly, viz. in pneumonic consolidation and in phthisis. It is modified bronchophony, and is probably due to the vibration of the collapsed bronchial tubes.

3. **Auscultation of the pleural sounds.**—In health one surface of the pleura glides over the other without a noise. In inflammation of the pleura as the surfaces become roughened from exudation, they rub or grate against one another, producing not only the pleural fremitus detected by palpation, but also various sounds to be appreciated by auscultation.

The sounds vary in loudness. They are rubbing to-and-fro sounds, sometimes heard throughout the whole of inspiration and expiration, at other times only at the end of inspiration, sometimes so rough, superficial, and characteristic as to be most easily diagnosed, at others faint, localised, and so like crepitation in the lungs as to be with difficulty distinguished from those sounds. When of the last-named character the term friction or pleural **crepitus** is sometimes used. Friction-sounds may sometimes be increased by pressure with the stethoscope over the affected part.

Succussion sound is heard when fluid and air are together present in the pleural cavity. On shaking the patient whilst auscultating his chest a splashing sound is heard, which may be imitated by shaking a china vessel half full of water. It is a sound almost characteristic of pyo-pneumothorax, but not quite, since it is occasionally heard over large vomicæ in the pulmonary substance.

Metallic echo, also called metallic or ringing râle, is really a special character added to the respiratory cough, or voice sounds in a large cavity containing air. A similar character attends the gurgling crepitation which is heard when there is fluid as well as air in it.

The **bell sound** is a clear metallic sound which is heard on listening over a large air-containing cavity, whilst a coin held against the chest is struck with another coin. The sound is particularly well marked in pneumothorax, but is also sometimes to be heard over a phthisical vomica if sufficiently large. A modification of the bell sound is as common if not more common under similar circumstances produced, one may suppose, by the same physical causes; it is a less clear resonant sound, like that produced by striking upon an iron anvil, and may be called the anvil sound.

4. **Auscultation of the heart.**—The position of the apex-beat marks the point on the surface of the chest where the first sound of the heart is to be heard most distinctly. The second sound is, however, loudest at the base over the situation of the semilunar valves. It is hardly necessary to do more than mention the characters of the heart-sounds. The first is longer than the second, duller and less clear, the second is a short sharp sound of higher pitch than the first. There is scarcely any interval of time appreciable between the two sounds, although such a one exists theoretically. There is, however, a distinct pause after the second sound.

As regards the **causes of the** heart-sounds it will be as well to arrange in order the cardiac and vascular events which are contemporaneous with each sound, before indicating those which may produce it.

Events which are taking place in the heart and **great vessels** during the time of the

First sound.—Systole of ventricles. Tension of auriculo-ventricular valves. Diastole of auricles, blood beginning to flow into auricles.

First pause.—Impulse of the heart against the chest wall. Entrance of the blood into the arteries.

Second sound.—Diastole of ventricles, **the auriculo-ventricular valves** being open. Continued **diastole of auricles.** Closure and tension of the semilunar valves. Recoil of aorta and pulmonary **artery.** Recoil of the heart.

Second pause.—First part, continued filling of the auricles and ventricles. Second part, systole of auricles. Complete distension of the ventricles, the auriculo-ventricular valves being open, and the semilunar valves closed.

Causes of the first sound.—Of the events above mentioned, which occur at the same time as the first sound, it is generally admitted now that those which are concerned in its production are the first mentioned, viz. **the muscular contraction**, which is much more **prolonged** than the **sudden** contraction of the skeletal muscle, and so produces a sound, even if the contraction of the heart be a single contraction, and not a tonic spasm, **which** some deny, and **the valvular sound**, produced by the vibration of the auriculo-ventricular valves, which are **already closed** when the ventricles begin to contract.

Causes of the second sound.—Experiment has proved beyond all shade of doubt that the second sound is due to the closure of the semilunar valves and their tension, caused by the bound of the blood against them on **diastole** of the ventricles. The vibration of other parts in their neighbourhood may perhaps augment the sound.

The time taken by each of the events of the cardiac cycle may be put in the form of a table thus:—Supposing the pulse $= 60$ per minute, each revolution would occupy one second.

Contraction of auricles would take $\frac{1}{10}$th second.
      ,,   of ventricles $\frac{4}{10}$ths   ,,
Pause $\frac{5}{10}$ths   ,,

Quiet period in which there is no contraction of auricle or ventricle $\frac{4}{10}$ths second.

The heart-sounds may be abnormal in the following ways. They may be—
1. **Too loud or too feeble.**
2. **Doubled or reduplicated.**
3. **Replaced, obscured, or augmented by murmurs.**

1. The sounds **vary in loudness** } In individuals and in the same individual at different times. Thus, exercise, or excitement may greatly **increase both sounds,** and in thin subjects they are heard much more distinctly than in the muscular **or fat, in whom** they are imperfectly **conducted** through the thickening over the chest. The sounds are louder in expiration than during **inspiration.** All these are physiological differences. General hypertrophy of the heart may increase both sounds; hypertrophy **of the** right ventricle increases the pulmonary second sound, and hypertrophy of the left ventricle increases the **aortic** second sound. Physiologically the sounds are **abnormally feeble** in stout persons. Pathologically a similiar condition is to be expected **when the heart acts feebly as in** chronic wasting **disorders and** fevers, **and** in anæmia, in **fatty** degeneration **of the heart** or atrophy—in pericarditis with effusion —in emphysema. The second aortic sound is sometimes very feeble in advanced mitral disease.

2. Either sound may be **doubled** or reduplicated. } When the first is doubled, the cause is said to be the want of synchronism between the vibrations of the segments of the auriculo-ventricular valves; it may be physiological or pathological, but in the latter case is not diagnostic of **any heart lesion.** When

the second sound is doubled, it must be taken to indicate that the aortic valves do not close at the same time, or that there is want of synchronism between the vibrations of the segments of one or both valves.

3. **Cardiac murmurs.** Any sound produced by the action of the heart which alters, replaces, or prolongs either of the normal heart-sounds constitutes a cardiac murmur. This does not include pericardial, pleuro-pericardial, pneumo-pericardial or cardio-pulmonary sounds, although they are produced by the heart's action.

The causes of cardiac murmurs are usually to be found at the valvular orifices. Here the conditions for the production of a **veine fluide**, the invariable cause, according to Chauveau, of the *bruit de souffle* are well established. A "veine fluide" is constantly formed whenever blood passes with a certain force from a narrowed into an actually or relatively dilated part of the circulatory system. He adds that no bruit can be produced by mere roughness of the coats of the vessels without an alteration of their calibre, or by alteration of the character of the circulatory fluid alone. It is obvious, however, that the conditions for the production of a "veine fluide" can be satisfied at other places besides the valvular orifices of the heart, and murmurs may not only be produced elsewhere in the heart, but also in the vessels.

The first division of cardiac murmurs proper, therefore, is into **valvular** and **non-valvular**.

1. **Valvular** murmurs are of two kinds, one produced by the orifice being constricted, absolutely or relatively to the cavity beyond, and the other produced by a portion of the blood being sent backward through a valve which is not competent to entirely

close the orifice. The former is called an obstructive or onward murmur, and the latter a regurgitant or backward murmur. According to Bergeon this latter is due to the valves projecting back into the wider part, so as to form a lip or rim with a circular cul-de-sac facing the current. A murmur which is thus produced, is audible on the proximal side of the obstruction, and is transmitted backwards or against the stream.

2. **Non-valvular murmurs** comprise those produced within the heart cavities, at the non-valvular orifices of vessels, and also in congenital diseases of the heart, patent foramen ovale, or incomplete septum ventriculorum.

Cardiac murmurs differ in i, **Character**; ii, **Time**; and iii, **Position** when heard loudest.

   i. **Character.**—They may be blowing, sawing, rasping, filing, whistling, cooing, crumpling, churning, or what not. They may be loud or soft, high pitched or low pitched, clear or dull. Some so faint as scarcely to be heard by the most careful auscultator, others so loud as to be heard all over the chest, obscuring other sounds, and possibly audible to the patient himself. It is needless to say that these characters are not diagnostic, neither is the intensity of the murmur any measure of the gravity of the lesion producing it. In fact, it not unfrequently happens that the loudest murmur is produced by the smallest lesion, and *vice versâ*. The intensity depends to a considerable extent upon the energy of the cardiac contraction, and a faint murmur may often be changed into a loud one by active exercise, as by walking briskly, or moving the arms up and down quickly; a murmur is generally plainer when the patient is lying down or sitting up than when he is standing; occasionally the reverse is the case. Musical murmurs are due to irregular vibration of the veine fluide, or to the vibration of a membranous substance suspended in the blood stream (most commonly) — Bergeon.

ii. **Time.**—Murmurs are either systolic or diastolic. A diastolic murmur occurring at the end of the second pause is called presystolic. Systolic murmurs are likely to be much louder than diastolic; they occur with the impulse of the heart. A systolic murmur produced at the mitral or tricuspid valve indicates incompetence, at the aortic or pulmonary orifice indicates obstruction and stenosis. A diastolic murmur proper produced at the aortic or pulmonary valve indicates incompetence, and a presystolic murmur at the mitral or tricuspid valve indicates obstruction or stenosis.

iii. **Position where heard loudest.**—Generally speaking at the part of the chest wall nearest to the site of their production. Thus, aortic murmurs are heard in the second interspace on the right side close to the sternum, and up and down that bone by conduction. Pulmonary murmurs are heard loudest on the left side in the second interspace close to the sternum. Tricuspid murmurs are heard best over the xiphoid cartilage and mitral at the apex beat in front, and at the angle of the scapula behind. The exact reason why mitral murmurs are always so loud in the latter situation is difficult to see, except that it is the spot on the chest wall where the sound is conducted *most directly*, although of course it is not that which is the nearest to the site of its production.

Hæmic murmurs.—In addition to the organic murmurs of the heart due to actual lesions, there are others which are not so produced, but which are due to an abnormal condition of the blood.

Let us now turn to these **inorganic, anæmic, or hæmic murmurs.** They are a variety of valvular murmurs, and are especially likely to occur in chlorotic girls, but also are noticed sometimes during the course of fevers, hæmorrhages, and sometimes in pregnancy. On the veine fluide theory they are thus explained by Chaveau. He says that in anæmia the mass of blood is diminished, and the smaller arteries, veins, capillaries, and the heart itself, accommodate themselves to this condition and contract, but the **aorta and pulmonary cannot contract to the same degree.** Their diameter remains, therefore, proportionately larger than that of the orifices through which the blood enters them, and the conditions for the *bruit de souffle* are fulfilled. The result is favoured by the greatly increased force with which the blood is driven into the arteries by the heart. This explanation would be sufficient supposing anæmic murmurs were heard equally well over the aorta and pulmonary valves, and never louder elsewhere. As a matter of fact, however, in the **pulmonary district** they are nearly always louder than over the aorta, and **sometimes** are certainly loudest at the apex. When the latter is the case it is probably due to a temporary incompetence of the mitral valve due to a relaxed support to the auriculo-ventricular ring, from tonelessness of the heart muscle itself, or to a degree of dilatation; but why the murmur is louder over the pulmonary orifice than over the aorta is not yet satisfactorily explained. Anæmic murmurs are, as regards

 (i) **Character.**—Soft, and blowing, never prolonged, and not permanent, not marked by increase on exercise or on lying down;

 (ii) **Time.**—Almost invariably systolic;

 (iii) **Position.**—Generally loudest over pulmonary **valves,** but sometimes equally so over aortic, and sometimes at the **apex-beat.** Rarely heard over all the valves.

(iv) **Accompanying conditions.**—The appearance of the patient, the general pallor, certainly aids the diagnosis, and the fact that there are no consecutive heart changes renders it probable that the murmur is **anæmic**, inorganic, and temporary.

Pericardial murmurs are friction sounds produced by the rubbing of one surface of the serous pericardium upon the other when roughening by inflammatory exudation has occurred. Similar sounds are also heard in the rarer cases of roughening of the same membrane by new growths, hæmorrhages, or thickenings. They present the following peculiarities in

(i) **Character.**—They are rough, grating to and fro, or rubbing and scratching sounds; very superficial; increased on pressure up to a certain point; **increased also by** the patient leaning forwards; altered, *i. e.* increased or diminished **by alteration of** position. Often accompanied **by a friction** fremitus. Not permanent.

(ii) **Time.**—They are heard independently of the heart sounds, but may be **systolic, diastolic,** or both, or even presystolic.

(iii) **Position where heard loudest.**—They are very localised, and not as a rule heard loudest at the situations where valvular murmurs are loudest, but are as a rule most marked at the **base, or to the** extreme left of the cardiac region, **and are** seldom heard out of the area of cardiac dulness.

Pleuro-pericardial and pneumo-pericardial murmurs are produced when the pericardium moves against the roughened pleura, or when the outside of the pericardium is roughened, and by moving against the lung produces a to-and-fro sound. These sounds are with difficulty distinguished from pericardial sounds proper.

Cardio-Pulmonary murmurs. — (1) A blowing systolic murmur is sometimes heard over lung-cavities and in their neighbourhood, produced by the contraction of the heart in the following way:

It is supposed that the air in the cavity is jerked out of it as it were by each systole of the heart, and that a systolic sound is thus produced, and that during the diastole the air returns, and its return may possibly be accompanied by a faint murmur. These murmurs may or may not be diminished by cessation of breathing. The movement of the heart may produce a murmur, generally systolic in time when the lung is fixed by pleural adhesions in front of the pericardium. A similar sound occurs on a deep inspiration under such circumstances.

2. A systolic murmur may be heard in the pulmonary artery when the apex of the left lung is consolidated, if there be at the same time contraction of its tissue. The contracting tissue presses upon the artery, and the conditions for the production of a bruit de souffle are set up. In a similar manner systolic murmurs may be produced in other parts of the lung by pressure upon the larger branches of the pulmonary artery.

3. A subclavian systolic murmur may be produced by the pressure of consolidated tissue of the apex of the lung or thickened pleura upon one subclavian artery. The murmur is increased by inspiration.

Arterial sounds and murmurs. — On auscultation of the subclavian and carotid vessels for some little distance from their origins the sounds of the heart, especially the second sound, may be distinctly heard. Under certain conditions murmurs too may be heard in these vessels. These murmurs are either transmitted, having been produced at the heart, or local, produced in the vessels themselves either naturally or by pressure of the stethoscope.

i. Transmitted arterial murmurs. — Mitral and aortic murmurs may be heard in the large vessels, but the aortic naturally the better. The systolic, obstructive, or **onward** aortic murmur, is the one which is most often to be made out, but **diastolic valvular** murmurs if loud are also heard distinctly; and sometimes not only in the large arteries, but even as far as the radials.

ii. **Locally produced murmurs.**—Of those which are **naturally** produced the most important is the aneurysmal murmur, which is either systolic or diastolic, but may be double. For the typical production of this murmur a sacculated aneurysm with a constricted orifice and a rapid stream of blood through it are required, yet natural murmurs may be produced in the arteries, by mere dilatation, by thickening, or by narrowing of their calibre from any cause.

Of the **artificial** pressure murmurs, the more common is systolic in time; and this may occur in healthy conditions when very slight pressure is exercised on the artery, but it is greatly exaggerated if from any cause the heart's action is excited or violent, as in hypertrophy of the left ventricle. Under these circumstances the murmur may be produced even in the smaller arteries.

The diastolic murmur is not so loud or so often heard as the systolic, and although it may be produced in such a vessel as the femoral artery under other conditions, *e. g.* aneurysm, chronic atheroma, hypertrophy of the left ventricle, &c., yet it is considered as generally indicating aortic regurgitation. It is due to the rebound of the blood past the point of obstruction during the ventricular diastole.

Venous murmurs.—The internal jugular vein of the right side at its junction with the subclavian to form the innominate is the situation in which venous murmurs are most likely to be heard. These murmurs are of two kinds, i, the **continuous murmur or hum**, and ii, the **intermittent murmur.**

i. **Continuous hum or bruit de diable.**—Especially

well marked in chlorotic females, in whom it is heard at the junction of the veins above mentioned, without any pressure on them with the stethoscope. It is a persistent hum, soft and blowing or whizzing, but at times roaring, whistling, or singing. It is quite without relation to the systole or diastole of the heart, but is continuous through both. It increases in intensity with increase in the rapidity of the circulation; is loudest during inspiration, and during the systole of the ventricles; is increased by turning the head to the opposite side, and diminished by lying down. A similar hum, not so loud however, may be produced in the healthy condition by pressure upon the veins with a stethoscope.

ii. **Intermittent venous murmurs**, præsystolic, systolic, or diastolic, have been occasionally heard. The systolic in tricuspid regurgitation. They are not very important.

There remain to be considered certain other aids to diagnosis in chest diseases which are not in such common use as the four methods hitherto studied. They are: i, **Stethometry**; ii, **Spirometry**; iii, **Pneumatometry**; iv, **Cardiography**; v, **Sphygmography**.

i. **Stethometry** consists in graphically recording the movements of the chest by means of an instrument,—a **stethometer** or **pneumograph**. There are several forms of the apparatus. A simpler form than the stethometer represented in the fig., consists of a thick india-rubber bag of an elliptical shape, about three inches long, to one end of which an india-rubber tube can be attached. This bag can be fixed at any required spot on the chest by means of a strap and buckle. Through the india-rubber tube the movements of the air in the bag are communicated to a tambour so arranged that its writing lever marks upon a revolving drum covered with paper. The alteration in the pressure within the bag can thus be communicated to the revolving surface. If two of these instruments be used at the same time on different parts of the chest the movements at those different parts can be carefully contrasted.

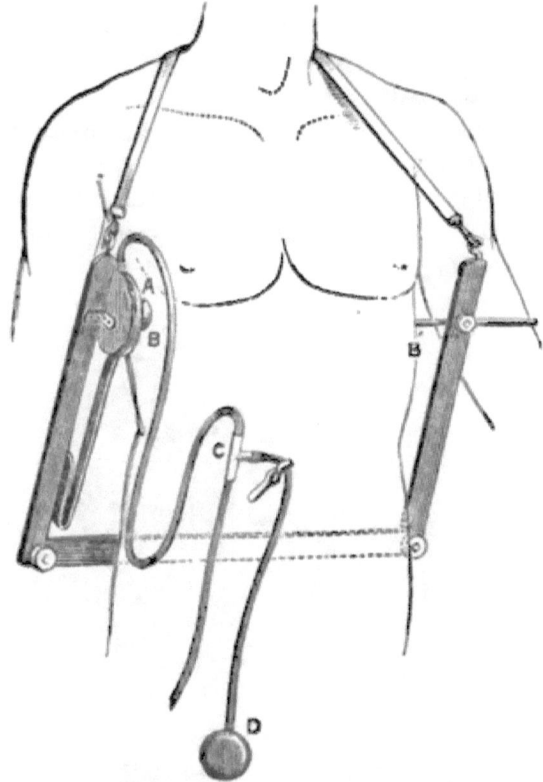

Fig. 17.—Recording Stethometer. A, Tympanum; B, ivory knob; B', rod which carries the knob opposed to B; C, T-tube, by which A communicates on the one hand with the recording tympanum, on the other with an elastic bag, B. The purpose of the bag is to enable the observer to vary the quantity of air in the cavity of the tympanum at will. The tube leading to it is closed by a clip when the instrument is in use.

It is by means of these graphic records that we know that expiration is really longer than inspiration, and that some parts of the chest move more than others during inspiration. As a diagnostic aid, it is unlikely to be frequently employed. In some diseases, no doubt, it shows very characteristic tracings, *e.g.* in emphysema, where the expiration is much prolonged. The respiratory tracing should, however, always be taken if possible simultaneously with one of the impulse of the heart as recorded with a cardiograph.

ii. **Spirometry.**—Is the method of measuring the amount of the vital capacity of an individual, that is to say, the amount of air which he can exhale from his lungs after the deepest possible inspiration. The spirometer is on the principle of a gasometer.

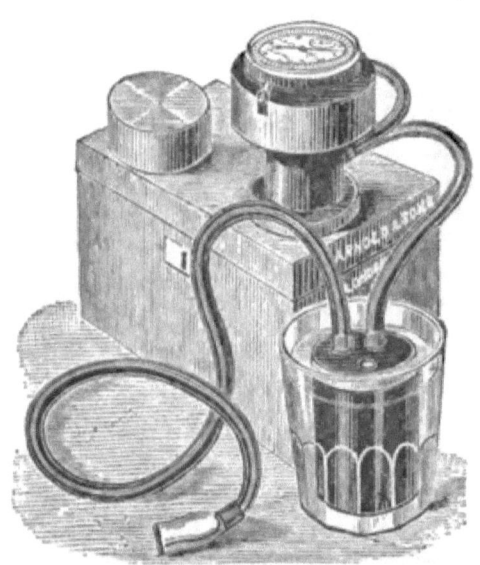

FIG. 18.—Spirometer (Lowndes').

The apparatus consists of two cylinders, one inside the other, the inner is open below, and fits into the outer one which is filled with water. It is properly balanced by means of weights which are attached to cords which work over pulleys at the sides, so that when the inner cylinder rises it remains stationary at its highest point. The height to which it rises is indicated by a scale showing the number of cubic inches of air which have entered it. The patient takes a very deep inspiration, and then expels by expiration as much air from his chest as possible through an india-rubber tube into the reservoir. The amount is then read off. Spirometry is not much used. Knowing the vital capacity of a patient in health, or at the beginning of a pulmonary affection of a chronic character, the changes in the amount of either increase or diminution might perhaps be some guide as to the progress of the case. Two forms of spirometer are shown in Figs. 18, 19.

iii. **Pneumatometry** is the method of measuring the force of inspiration and expiration by means of a mercurial manometer. The patient is made to breathe steadily through an india-rubber tube connected with the horizontal arm of one of the limbs of the manometer, which is half filled with mercury; the fall of the mercury in the arm nearest the patient during expiration, and the rise during inspiration, are measured by a scale attached to the instrument. It is found that the force put forth in expiration is greater than that of inspiration, the difference being from 20 to

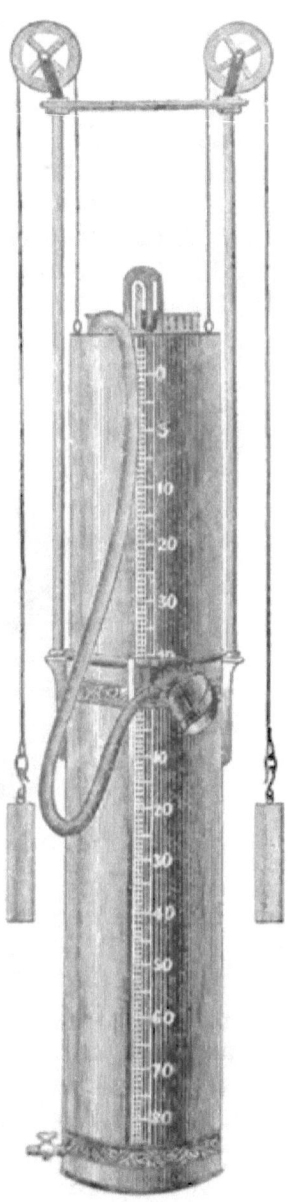

FIG. 19.—Spirometer (modified from Hutchinson's).

30 mm. of mercury. By the use of this instrument it is demonstrated that the force of inspiration may be lessened, with or without that of expiration, in certain diseases, *e. g.* phthisis, laryngeal obstruction, and pneumonia. In other affections the expiratory power is diminished, *e. g.* in emphysema, bronchitis, asthma, and mitral disease.

iv. **Cardiography** is the method of recording graphically the movements of the heart against the chest wall. This is done by means of a cardiograph. There are many varieties of this instrument. The one

FIG. 20.—Sanderson's Cardiograph.

which is most commonly employed is that of Burdon Sanderson. It is a modification of the original cardiograph of Marey. It consists of a metal cup closed in with india rubber, upon the centre of which is fixed a projecting button. The interior of this air-tight tympanum is connected by means of a piece of tubing with a small, similarly formed tympanum or tambour (Fig. 23), to which is attached a writing lever to mark upon a revolving drum. The button is made to press lightly upon the apex-beat, the movement of which is thus communicated from the first tympanum to the tambour, and so to the recording surface.

The cardiograph has been so far a physiological rather than a clinical instrument, but there is good reason to hope that it may be in the future of considerable help in diagnosis. By its aid may be made out at any rate the relation of the duration of the systole of the ventricles to their diastole, and consequently the exact effect of drugs which act upon the heart.

v. **Sphygmography** is the method of recording graphically the characters of the pulse, the examina-

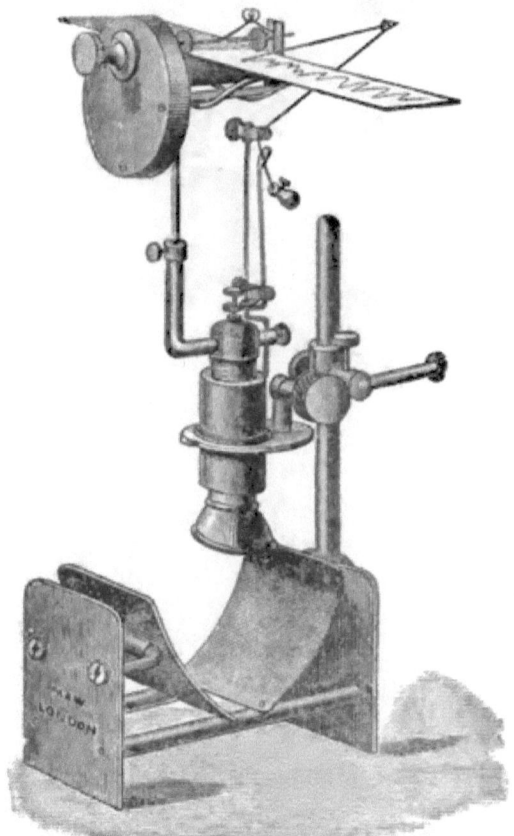

FIG. 21.—Pond's Sphygmograph.

tion of which is treated of in the next chapter, by means of the sphygmograph. That instrument is too

120 THE PHYSICAL EXAMINATION OF THE CHEST

well known to need any description. The varieties employed generally are Sanderson's, or Mahomed's modifications of Marey's sphygmograph (Fig. 22).

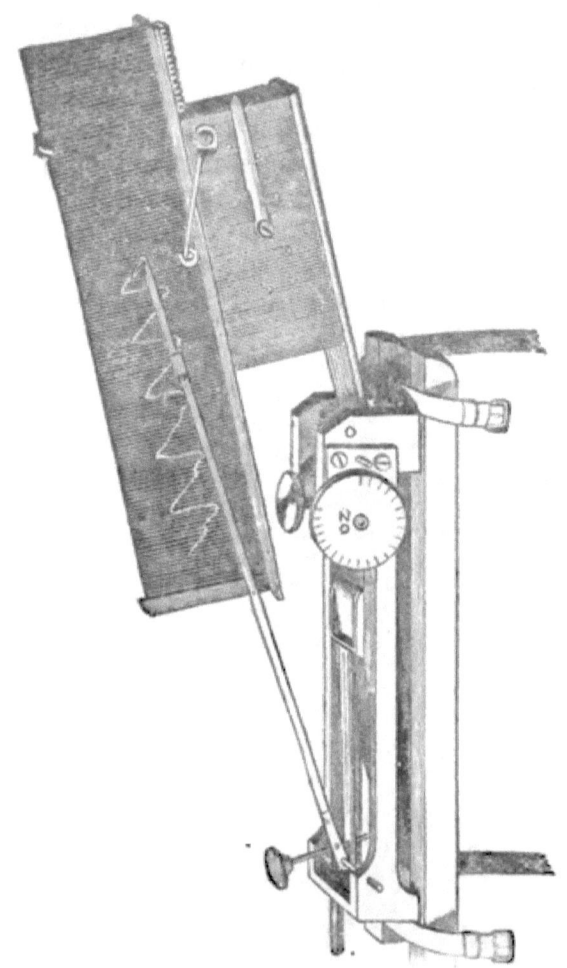

FIG. 22.—Mahomed's Sphygmograph.

Pond's sphygmograph (Fig. 21), or Dudgeon's modification of it, is easy to work, but is stated to be not so reliable as Marey's.

In a normal pulse-tracing there are several well-

USE OF THE SPHYGMOGRAPH

known features. First, there is the sudden rise of the lever, and the more gradual fall; as the lever falls it is raised slightly several times, and especially about half way in its descent to form the dicrotic notch produced by the so-called dicrotic wave, the other elevations of the lever being called the pre-dicrotic and post-dicrotic respectively. As a clinical instrument the sphygmograph has received much attention, but it is doubtful whether it is ever of much diagnostic value in cases where a careful examination of the pulse with fingers fails to indicate its characters. It is, however, useful in accentuating and confirming the evidence obtained by digital examination. The characters of the pulse will be considered in the next chapter, and sphygmographic tracings of it in certain conditions will be found in the chapter on Valvular Diseases of the Heart.

FIG. 23.—Marey's Tympanum and Writing Lever. To the right is the tube connected by some unyielding elastic tubing with the cardiograph.

CHAPTER IV

THE PULSE AND ITS INDICATIONS

THE application of the finger to the radial artery, or "feeling the pulse" as it is called, is a method of physical examination employed universally. It is no doubt true that it is a method less esteemed at present as a means of diagnosis than it was formerly. This is not surprising when we consider how many other helps to diagnosis have been introduced of late years. Nevertheless a careful study of the characters of the pulse cannot be too much insisted upon, because from them, if rightly interpreted, we obtain the chief indications of the condition of arterial tension, as well as of the factors upon which the blood pressure depends. These factors, which have been touched upon in the first chapter, are :—1, **The force and frequency of the contractions of the heart**; 2, **The condition of the arterial coats**; 3, **The condition of the peripheral resistance**, and our examination of the state of the artery ought to give us information both as to each of them, as well as of their resultant, *i. e.* the state of the tension of the artery.

The pulse is usually described as an expansion of the artery produced by a wave of blood set in motion by the injection of blood by the ventricular systole into the aorta. Dr Broadbent says, however, that this is scarcely accurate, or that at any rate this is not what we feel when we put our fingers upon the vessel. He goes on to say, "To feel the pulsation in an artery, or to take a sphygmographic tracing, a certain degree of pressure must be applied to the vessel, and, as is well known, there must be a bone behind it, against which it can be compressed. What happens then is as follows: In the intervals between the pulsations, when the resistance by the contained blood is at its lowest, the tube of the artery is more or less flattened, then comes the so-called wave of blood propelled by the systole of the left ventricle, or, to speak more accurately, the liquid pressure in the vessel is increased, and this forces the artery back into its circular form. It is this change of shape from the flattened .... to the round cylindrical shape .... which constitutes for us the pulse." On consideration of Dr Broadbent's description we are bound to admit that it is more accurate than that which is ordinarily given as regards the pulse we feel in the radial, although the visible pulsation of that or any other artery must be an expansion or increase in calibre, due to the rise of blood pressure at each ventricular systole. When we feel the pulse, then, we endeavour to discover from our examination the **frequency, degree, and duration of the rise of blood pressure**, as well as the **condition of the tension of the artery between each such rise** and whether the rise is **regular in time, and equal in amount.** We also gauge the **dicrotism.**

These are the qualities then which it is necessary to consider separately.

   a. **Frequency of the pulse.** The frequency of the pulse differs at different ages as is well known. It is very rapid in infants, beating 130—140 times a minute, and during the first year of life 115—130 times. After this the beats of the pulse gradually diminish in number per minute, until they reach the average of 70—80 in adult life. In old age, the number is from 60—70. In decrepitude, the beats may become a little faster.

The frequency of the pulse depends upon the number of the heart beats, and so it may be called one of the purely cardiac qualities of the pulse. The number of beats may, however, either be diminished below or increased above the normal.

A. **Infrequent.**—The number of beats of an infrequent pulse may be taken as below 40 per minute. A pulse of 60 is not rare, but we seldom meet with instances of pulses below that number. Infrequent pulses may be either

a. Physiological.—Some individuals with pulses beating very infrequently, enjoy perfect health.

Or *b. Pathological.*—The rapidity of the pulse is diminished in jaundice, diphtheria, cerebral hæmorrhage, albuminuria, and convalescence from acute diseases.

B. **Too frequent.**—The pulse may be considered too frequent if above 80—100. Sometimes may be as high as 200 or more.

a. Physiological.—Some pulses habitually and normally exceed the average number of beats.

b. Pathological.—Any of the conditions already enumerated which produce palpitation, especially nervous causes, will produce a too frequent pulse, as will also any condition which relaxes the arterioles and diminishes the peripheral resistance.

The chief causes of frequent pulse apart, from those mentioned as causing palpitation, are febrile or nervous excitement, hysteria, and Graves' disease.

b. **Rhythm of the pulse.** The natural rhythm of the pulse may be altered, so that occasionally a pulse beat may occur every now and then at an irregular interval, every beat being separated by equal increments of time when the pulse is regular. Such a pulse is **irregular in time**. If a beat now and then be smaller than its fellows, the pulse is irregular **in volume**, and so on. The **former** abnormality may become more and more marked, so that the irregularity may occur oftener until there results a pulse which has lost all trace of regularity. Again, slight irregularity in volume and force may be exaggerated so that the beats run in pairs, with a longish interval between each pair, the second beat of each pair being weaker than the first; such a pulse is called the **pulsus bigeminus**, the cause of which is not known; or three beats may be followed by a pause in the pulsus trigeminus. A form of pulse, too, is described in which a good beat is followed by a series of gradually diminishing beats until a second good beat occurs.

When a beat of the heart fails to be represented in the radial pulse, the beat is abortive, and an intermission occurs. A pulse may be rhythmically or not rhythmically irregular, and so may or may not be rhythmically intermittent. Sometimes intermissions are due to interruptions of single heart beats.

The most frequent causes of irregularities and intermissions of the pulse are: (*a*) **Valvular heart disease**, especially mitral valve disease. (*b*) **Dilatation** from any cause; (*c*) **Fatty degeneration.** (*d*) **Pericardial effusion.** (*e*) **Mechanical interference** with the heart's action. (*f*) **Dyspepsia.** (*g*) **Action of drugs**, *e. g.* excessive doses of digitalis, tobacco, &c.

It should be recollected that the pulse in the radial on the side of an aneurysm implicating the innominate artery may be a little later and smaller than that in the other radial.

c. **Volume of the pulse.** The volume of the pulse varies:
i. Directly with the capacity of the left ventricle and the **amount of blood** discharged with each beat.

ii. Inversely with the time occupied by the ventricular systole.

iii. Directly with the capacity of the artery.

iv. Inversely with the degree of arterial tension at the beginning of ventricular systole depending upon: (*a*) The amount of blood contained in the **arteries.** (*b*) **Tonicity** of the muscular coat.

e. g. The volume of the pulse is **increased** in aortic insufficiency and diminished in aortic stenosis. Sometimes the pulse is so slight as to be **thready** or **tremulous** merely.

Pulsus paradoxus is a rhythmical variation in the volume of the pulse produced by respiration, the beats being diminished during each inspiration, or altogether ceasing if the inspiration be deep. The conditions predisposing to this form are anything which interferes with the discharge of blood into the aorta, or which **obstructs the entrance of air into the lungs.**

d. **Tension of the artery.** The tension of the artery includes two conditions, first, the minimum tension, and second, the maximum. The minimum tension is the condition immediately before, and the maximum tension the condition at the beat.

As before described in Chapter I, the **nervous arrangement of the** circulation is such that, as a rule, if increased arterial tension be due to increased peripheral **resistance,** the pulse-rate is lowered in consequence of the stimulation of the cardio-inhibitory centre by the increased pressure, and *vice versâ* ; but if increased pressure be due to the heart, the pulse-rate rises with the **rise** in the **blood pressure.** If the capillaries and arterioles be **relaxed, the** minimum tension **is** small, and **the** pulse is easily **compressible** or **soft.** Under such conditions it is possible to get pulsation in the veins. The **amount of** maximum tension depends **upon other** things, viz. upon the size of **the artery** and upon **the action of the** heart. **If the artery** be **small and** the **heart's action weak, we have a small and weak** pulse. **If the artery be large and** the **heart's action strong, we have a** large **wave** and **strong but** short **pulse. If the** artery be large and the heart's action **weak, we have a** large, short, and weak pulse.

Conditions of low tension:—(1) *Physiological.*—A pulse of low tension may be an individual or family peculiarity.

2. *Pathological.*—i. Fever. ii. Cardiac degeneration and debility. iii. **Anæmia** (sometimes).

High tension.—The pulse of high tension is one in which **the minimum tension is exaggerated, so that the artery** cannot be easily **compressed between the beats,** the wave is **long and sustained, and the** pulsation is more **felt** as the pressure **is** up to a certain **point increased.** These features are especially marked when the artery is large from prolonged distension, but it is possible to have high tension with a small artery.

High tension is usually due to an increase in the **peripheral resistance, either** in the **arterioles (in** pyrexia, **cerebral** and spinal diseases, **and** fever) or in **the capillaries (in** renal disease and allied conditions), **and with** high tension, therefore, **we** generally have **a** forcible and infrequent action of the **heart.** Sometimes **with** this high tension the heart may not contract forcibly; if this is the case one would expect high tension of the arteries to continue, but only for a very short time.

Conditions of high tension:—1. *Physiological.*—A high-tension pulse may be an individual or family peculiarity.

2. *Pathological.*—i. **Kidney disease**, especially granular and contracted. ii. **Gout** and **Lithæmia**. iii. **Lead poisoning**. iv. **Anæmia** (very often), and its converse **plethora**. v. **Pregnancy**. vi. **Constipation**. vii. **Chronic bronchitis and emphysema**.

e. **Duration or rapidity of the pulse.** } This quality is closely connected with the tension of the vessel. By it we intend to express the relation of the duration of the rise in blood pressure to its fall. If the rise is sudden and short we have a **quick** pulse, or in a more exaggerated form a **bounding** pulse. If the rise in pressure is prolonged, as in over-full or in inelastic vessels, the pulse is **slow or sluggish**.

The best example of a pulse in which the rise is sudden, or sudden and large but of very short duration, is seen in aortic incompetence, where the injection of the blood into the arteries is forcible owing to an hypertrophied left ventricle, but the increased blood pressure is speedily diminished by the exit of blood from the arteries through two outlets, viz. outwards into the capillaries and backwards into the heart.

f. **Dicrotism of the pulse.** } This quality of the pulse when marked must be considered to be merely an exaggeration of the dicrotic wave which is seen to exist in most pulses when a sphygmographic tracing is examined. It is a slight and temporary rise in the pressure of the artery more or less midway between the maximum and minimum of tension. It is almost certainly a recoil wave from the closed and stretched aortic valves. A pulse would be called **dicrotic** only when this recoil wave is appreciable by the fingers. It occurs particularly in fever and in other conditions of a low-tensioned artery.

CHAPTER V

DISEASES OF THE PLEURA

In considering in order the diseases of the thoracic viscera, it seems convenient to begin with an account of the affections of the serous investment of the lungs, i. e. of the pleura, before we enter upon the description of the diseases of the lung substance itself, and of those of the respiratory mucous membrane. The present chapter, therefore will be devoted to the subject of the **diseases of the pleura.**

I. Inflammation of the Pleura, Pleuritis or Pleurisy

The pleura, on each side, it will be remembered, forms a complete and independent closed sac, the walls of which are practically in contact, which is interposed between the chest walls and the lung. By its outer surface it is attached to the former, forming the **parietal** pleura, and by its inner surface it is attached to the lung, and forms a **visceral** serous investment for it. The external surface of the visceral or pulmonary pleura is able to glide without friction over the internal surface of the parietal pleura, as they are both beautifully smooth and polished, and thus the movements of the lung in respiration can take place in health without hindrance, just sufficient fluid to moisten the interior being found within the sac.

When the pleura becomes the subject of inflammation, the exudation takes place upon the opposed surfaces, and the effects are produced chiefly within the sac; thickening by exudation into the few layers which form the serous membrane itself, although generally present, being an occurrence of secondary importance.

Conditions under which it arises.—The pleura may become inflamed under a variety of different conditions, but the part which the inflammation of the membrane plays in each is not the same. It is sometimes an independent disease attended with constitutional as well as local symptoms, at other times it is a local affection attended with symptoms almost entirely local, and again it is not infrequently a complication arising in the course of other affections, especially of the specific fevers. In the last case it is attended, it may be, with no symptoms of its own. It is convenient to discuss these several conditions in order. Pleurisy may occur:

(1) **After exposure to wet and cold.** A history of exposure to wet or cold in some form immediately preceding the onset of the attack is generally to be obtained when pleurisy arises as a distinct and independent or idiopathic affection. A sudden change from a warm into a cold temperature, sitting or standing in a draught of cold air, sitting in wet clothes or with cold feet, exposing the body to cold after violent exercise, a damp bed, in short any condition in which by cold or moisture the heat of the body is abstracted more rapidly than it is generated from within, unless such loss of heat be replaced by exercise or cordials, as Watson says, may derange the balance of the circulation and set up pleurisy. Generally speaking the action of cold must be prolonged and not momentary, and also it is to be noticed that harm is most likely to happen if the person who is exposed to this possible source of danger is weak, over-tired, or in some other way not up to the standard of health. Sleeping whilst exposed to cold and damp, and special delicacy

of constitution, should also be mentioned in this connection. Why pleurisy should arise in one person, bronchitis in another, pneumonia in a third, or Bright's disease in a fourth, from a cause to all appearance exactly similar, or why either affection may occur in the same person at different times from similar exposure, is a question which has been often discussed with but little result. It is, **however, a fact that** such anomalies do occur. It is possible that the variety of pleurisy which arises under such circumstances is a specific disease, but there is no evidence at present which proves this supposition.

(2) From **extension of inflammation** from neighbouring organs, especially from the lungs and pericardium.

Pleurisy is almost certain to occur with acute croupous pneumonia, and generally as a part of the disease, pleuro-pneumonia, but sometimes possibly it may be merely set up by the irritation of pneumonic consolidation in its immediate neighbourhood.

When pleurisy arises in connection with the next three conditions, it must be looked upon as a *local* affection; it is also a local affection, at any rate sometimes in (2).

(3) From irritation by new growths, *e. g.* by carcinoma; by consolidated patches in the lung; by tubercle or by aneurysm of the aorta.

(4) From **traumatic cases**, as from after blows, stabs, or injuries of any kind to the chest, including gunshot wounds. Of these accidents by far the most common, among civilians, is the injury produced by a broken rib.

(5) From **the entrance of pus** into the pleural sac, as in the case of rupture of an abscess of the liver, or of other organs into it; or of blood, as by rupture of an aneurysm or intercostal vessel;

or of air; or of the **contents** of phthisical **cavities.**

In the next set of conditions pleurisy arises as a complication, and is *symptomatic* of **the** affections.

(6) **As a complication.**—During the course of septicæmia, puerperal fever, scarlet fever, typhus fever, diphtheria, and other acute specific fevers, and in Bright's disease of the kidney both acute and chronic, and in acute rheumatism associated with pericarditis.

From a consideration of the above very numerous and extensive conditions under which inflammation of the pleura may arise, more numerous than in the case of other serous membranes, a proneness **to** inflammation of that serous membrane must **be** inferred, more marked than in other similar structures. We must look **to the** peculiar condition which the pleura and its blood-vessels occupy with respect to atmospheric pressure for an explanation.

Pleurisy may attack persons at any age and of either sex, men more often than women, and rather **more** frequently appears on the left than on the right side.

Pathology and varieties.—When from any cause inflammation is set up in the pleuræ, either by local irritation or general infection, the blood-vessels become either locally or generally engorged with blood, and an exudation takes place. The endothelium covering the membrane becomes cloudy and proliferates. The exudation may **extend** into the **scanty subserous tissue, but appears more** particularly upon the **inner surface of the serous membrane** as a false membrane which varies greatly in thickness, and is made up of fibrin, colourless corpuscles which have escaped from the blood-vessels, prolife**rated** endothelium, and proliferated connective-tissue corpuscles of the pleura itself. Together with **the** deposit upon the surface there is more or less fluid exudation into the sac of the pleura. The inflammation of the tissue begins in one or more spots, and it may either rapidly extend until the whole membrane is implicated, or it may continue to be localised from first to last.

The classification of the pathological and clinical varieties of pleurisy is based upon the nature and quantity of the inflammatory exudation.

Four varieties are thus to be distinguished, viz.:

1. **Dry pleurisy** or pleuritis sicca. This form is essentially local, and arises at the seat of some definite point of irritation, as, for example, a consolidated patch of lung tissue beneath the serous membrane. It is very common in phthisis, and is one of the chief causes of the pains which are so often complained of in that and other chronic chest diseases. There is slight, if any, exudation, but roughening of the surface of the pleura may be produced, and local adhesions may occur by the growing together of the two surfaces, which are vascular and covered with proliferated endothelium.

2. **Pleurisy with thick fibrinous exudation.** This form of the affection is especially likely to arise in connection with pneumonia, or with peritonitis and pericarditis, or from extension of inflammation from other neighbouring parts. In such cases the exudation is more abundant than in the first form. It is chiefly thick and fibrinous, but there still is associated with it more or less fluid, free in the sac. The part of the exudation deposited upon the pulmonary and costal pleura forms a false membrane in which blood-vessels speedily develop, and which is capable of quickly undergoing fibrillation. The surface of the membrane is rough and shaggy, and villous outgrowths can be seen projecting from it; it varies from half an inch to one or two inches in thickness. The

villous processes probably grow together, and form adhesions of various width, length, and extent. These adhesions are elongated into irregular cords, which form bands connecting the two surfaces when fluid separates the walls of the pleural sac. They may be of considerable extent. They are generally attached obliquely from surface to surface.

3. **Pleurisy with considerable serous effusion.** There is an insensible gradation between this and the preceding variety of the affection, but the extremes of both are distinct enough. Considerable serous effusion occurs as a rule in pleurisy arising from exposure to wet and cold, and less frequently in other cases, as, for example, in tuberculosis of the pleura. The effusion takes place rapidly, and may in a short time amount to several quarts. It may be either pale straw-coloured, clear and without sediment, or it may be cloudy, milky, curdy, flaky, greenish yellow, depositing a distinct sediment on standing. At the same time the surfaces of the pleura are covered with a more or less thick layer of uneven exudation, such as, but thinner than, that described in the second variety, and in many cases trabecular connecting bands traverse the fluid. The aspect of this false membrane is often distinctly greenish. The *fluid* removed from a pleurisy by tapping is, as a rule, serous, and closely corresponds with diluted liquor sanguinis, but the fibrin generating constituents are not generally in the proper proportion to produce a clot. A partial clotting, indeed, has already taken place from the exudation in the deposit upon the surface, but yet the fluid contains some fibrinogen. Sometimes, however, the fluid sets into a firm jelly as soon as it is removed, and in other rarer cases it clots as it is being evacuated.

The rupture of the newly-developed and thin-walled blood-vessels, which form in the false membrane either in the second or third variety of pleurisy, may complicate the appearance of the fluid or solid exudation by the introduction of more or less blood.*

The effects of the effusion upon the chest and its contents may be summed up as follows:—The side upon which the effusion has taken place is greatly distended, and the intercostal spaces may be stretched to their fullest extent, and may be bulged outwards; the diaphragm may be depressed or flattened. The heart is pushed over to the side opposite to the effusion (especially to the right). The lung on the affected side is compressed against the posterior wall of the chest, is practically airless and solid, of a colour varying from pale red to slate or lead colour, of a tough, leathery consistence, and coated externally with a thick fibrinous deposit. Under certain conditions *effusions may become encapsuled* by adhesions, and in this case, and when the effusion is partial, the lung suffers compression only in the part corresponding to the effusion. Under favorable circumstances, the liquid part of the effusion may become absorbed, and the fibrinous deposit, after undergoing fatty degeneration, clears up, the lung again becomes pervious, and nearly complete resolution is accomplished. In order that this satisfactory state of affairs may occur, it is essential that the effusion should be early and rapidly absorbed. Generally, however, adhesions take place between the walls of the pleural sac, and the lung is bound down to a

* **Analyses of Pleural Fluids.**—Several accurate analyses of pleural effusions have been made. In pleural exudation obtained by paracentesis in 1000 parts, according to Scherer, 64·48 were solids, of which albumen (native-albumins and globulins) constituted 49·77; fibrin (or fibrinogen) 0·62; extractives, 5·6; inorganic salts, 7·93. Hoppe-Seyler's analysis of the fluid from a hydrothorax showed only 42·41 solids in 1000 parts.

greater or less extent. The pulmonary, as well as the parietal pleura, continues to be thickened.

The thickened pleura is found to be made up of broad bands of fibrous tissue with connective-tissue corpuscles, and a similar fibrous change invades the subpleural tissue and lung. This invasion of the connective tissue of the neighbouring parts is often very extensive. It is sometimes particularly exaggerated in the thoracic walls, which become once and a half or twice as thick as natural, but it also occurs in the interalveolar and interlobular pulmonary tissue and elsewhere. This gradual implication of the lung in cases of chonic pleurisy constitutes one form of the so-called interstitial pneumonia.

The lung scarcely ever entirely resumes its former condition, being less expansible and elastic. In chronic cases the area of the affected side has a tendency to diminish, the chest walls fall in, the intercostal spaces are narrowed, the diaphragm is drawn up, and the heart is drawn by adhesions towards the unexpanded lung, the shoulder drops, the side moves imperfectly in inspiration, and the spinal column may become more or less curved towards the affected side.

It must not be forgotten that pleural inflammation, if sufficiently severe, may set up inflammation in the neighbouring parts, the lung, pericardium, peritoneum, &c.

4. Pleurisy with purulent effusion—Empyema—Pyothorax

In this variety of pleurisy, the formation of pus takes place, either primarily from the irritation setting up the inflammation being excessive, in which case it must be considered to be a stage beyond what occurs in the sero-fibrinous variety of the affection, or as occurring in a vitiated condition of the system, or secondarily in cases in which the effusion has been first of all of a serous or fibrinous kind, and from entrance of air or what not, the effusion gradually becomes purulent. There is nearly always a more or less thick deposit upon the lung.

The purulent effusion is said sometimes to undergo absorption after fatty degeneration of the solid constituents has taken place, but this must be very rare. The empyema may, if allowed to go on undisturbed for some time, tend to point (i) through the thoracic walls as high up in the front of the chest as the second interspace near the sternum, or even above this situation, or as low down behind as the pleura extends, or by tracking through the muscles it may open almost anywhere upon the surface. Or (ii) it may perforate the lung and be discharged through the bronchi, or (iii) it may find its way into the neighbouring serous cavities, the pericardium, or the peritoneum.

Symptoms and course.—The symptoms of pleurisy, as may have been inferred from what has already been said, vary with the nature and extent of the affection. The inflammation, if slight, may give rise to no marked disturbance, local or constitutional, or possibly to a pain, severe while it lasts, in the affected side only. In the independent disease which arises after exposure to wet or cold, however, the following symptoms are usually present in a greater or less degree.

i. **Pain or stitch in the side.**—A pain which comes on suddenly, a longer or shorter time after the exposure to the exciting cause is nearly always present. It is increased on taking a deep breath, on cough, or on any movement; it is also increased on pressure, unless it be firm and sustained, when the reverse may be the case. In character it is acute and stabbing, often so severe as to make the patient gasp, groan, or cry out. It is usually possible for the patient to indicate an exact spot where the pain is most intense, from which it radiates for a greater or less distance; sometimes it is felt at some distance from the seat of inflammation, sometimes in the shoulder, in the axilla, beneath the clavicle, or along the sternum, and when the diaphragmatic pleura is inflamed over the cartilage of the false ribs. As the effusion increases, the **pain abates**.

ii. **Dyspnœa**.—The respirations are rapid and shallow (25 to 30 or more). As a deep inspiration causes much **pain on the** affected side, the patient endeavours **to avoid the** necessity by breathing quickly, taking shallow breaths, and leaning towards that side. So from the first there is considerable increase in the rapidity of respiration. Later on the presence of a large amount of fluid on one side of **the chest** produces pressure on one lung, and consequent increased action of the other; in severe cases, both lungs may also be compressed. No doubt the dyspnœa is sometimes partly to be attributed to the accompanying fever.

iii. **Dry cough** frequently occurs, and may be very irritating and distressing, greatly aggravating the pain. There **may be scanty** frothy expectoration.

iv. **Fever**.—There is **generally some feverishness, varying greatly, however, in intensity**. Sometimes the pyrexia is ushered in with a distinct rigor or series of rigors, but although in certain cases the constitutional disturbances are considerable, being marked by full and frequent pulse, high temperature, headache, pains in back and loins, thirst, and furred tongue, yet as **a rule both the fever and the** constitutional **symptoms are slight**, the temperature seldom rising above 101—102°. Certain cases appear to terminate, as far as the fever is concerned, by distinct crises, about the end of the first week.

v. **Restricted decubitus**.—In few cases of pleurisy is the patient able to lie in any position indifferently; **at first he cannot lie on the affected** side, because the **pain** is thereby increased, and so he lies on his back or on the other side; later on, when effusion has taken place, he lies on or towards the affected side, in order that the sound lung may not be oppressed by the weight of the fluid upon it.

It is very remarkable that patients not seldom are met with among out-patients who on examination are found to have large effusions, and yet seem to have suffered little inconvenience, and have walked to the hospital.

vi. **Alterations in the shape of the chest**, which will be referred to below under the head of the physical signs of the affection.

As regards the **course** of the affection, when the pleurisy is of the dry variety the pain and dyspnœa are very temporary, although it should be remembered that while it lasts the former may be very intense and distressing; this is especially the case when the diaphragmatic pleura is inflamed (p. 162). We have frequently to do with pain in the side among out-patients at hospitals, with scarcely any physical signs, which is, we believe, more often pleuritic in its cause than is generally thought, and slight adhesions of the pleura very frequently found here and there in the chests of those otherwise healthy, when examined after death, have probably occurred during attacks which have been called dyspepsia or pleurodynia. Watson was disposed to agree with Cruveilhier in thinking that pleurodynia was nothing else in many cases than adhesive pleurisy.

When there is effusion the course may be acute throughout, beginning as above mentioned, with considerable fever, and terminating by crisis, or it may be acute to begin with, becoming chronic afterwards, or it may never be otherwise than chronic. In any case the exudation remains, and the length of time this takes to undergo absorption is very variable; sometimes so short a time as three or four weeks; in other cases, months; but there is no doubt that complete absorption does often take place, and that air freely enters the compressed lung. It is noticeable that the process of absorption is frequently intermittent, alternating with fresh exudations. When the effusion is very long undergoing absorption the constitutional symptoms are marked, a raised temperature continues, and the pulse remains rapid but becomes feeble, the patient wastes and becomes very anæmic. All these signs are the more marked if the effusion has become purulent.

Symptoms pointing to empyema.—Unfortunately there are no *certain* indications of the presence of pus in the chest until later on in the affection, but if

Rigors or repeated shiverings occur in the course of a pleuritic inflammation and are followed by prolonged

Hectic fever, the temperature undergoing a distinct evening rise, together with

Sleep sweats, marked emaciation, harsh skin, and clubbing fingers, we may assume that the effusion has become purulent. This is rendered certain when the empyema

Tends to point, and the skin in any particular part of the chest wall, especially high up and in front, gets red, shining, œdematous, and gives the sense of fluctuation under the finger.

Physical signs.

Of pleurisy with effusion.

I. On inspection (*a*) the intercostal spaces are flattened out if the effusion be very great or even prominent and distended.

(*b*) The affected side is dilated. This is confirmed by taking the measure of it with a tape measure, and comparing it with the sound side. It will be found to be a half to two inches greater than the sound side. A cyrtometric tracing may also be taken and measured.

(*c*) Absence of movement of the affected side.

142   DISEASES OF THE PLEURA

(d) Displacement of the heart's apex-beat away from affected side. This is especially the case when the left side has been attacked, as the apex-beat is then often to be felt to the right of the sternum.

II. **On palpation** (a) friction fremitus may be felt on the application of the hand, sometimes early in the affection as well as later on, when the fluid is being absorbed.

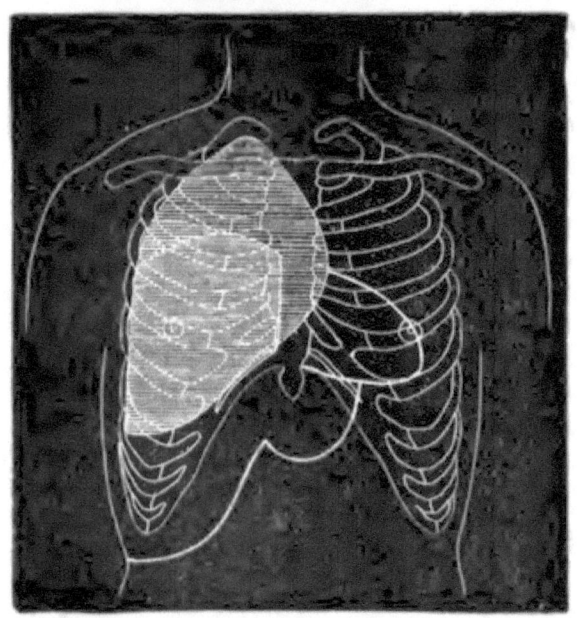

FIG. 24.—Diagram of the extent of the percussion dulness in right pleural effusion. The lighter shading indicates the dulness when the effusion is excessive, and the deeper shading when it is moderate.

(b) Vocal fremitus is absent or much impaired over the area of the fluid, and rather more marked than usual on the opposite side.

(c) The heart's apex-beat is found to be displaced to the side opposite the effusion, and the liver is felt below the ribs when the right side is affected (Fig. 24).

III. On percussion, there is dulness, absolute and resisting, to the upper limit of the fluid, which *occasionally* alters in level on alteration of position of patient.

The upper limit of the dulness is sometimes a straight line. Although it might be expected that this would generally hold good, and indeed such is stated by good authorities, yet, as a matter of fact, the dulness, both in front and behind, is bounded above by a curved line, at any rate when the effusion is moderate, when it is increasing, and also when it is diminishing. The highest part, both of the anterior and of the posterior curve is in the axillary region. The dulness may extend beyond the midsternal end. There is nearly always behind, corresponding with the compressed lung, a semi-tympanitic note, and the upper part of the chest is frequently quite tympanitic or hyper-resonant on percussion, as is also the opposite side, and my experience appears to justify the belief that the dulness tends to increase laterally even to the other side of the sternum before the upper part of the chest becomes dull.

IV. On auscultation (*a*) friction-sound is heard on inspiration or on expiration or on both, early or late in the affection. This may be so loud as to be heard without placing the ear on the chest, and without the aid of a stethoscope. (*b*) Very faint vesicular breathing, distant tubular breathing, or absence of breath-sounds over the seat of effusion, and sometimes very loud tubular breathing over a large effusion. Puerile breathing over the uncompressed part of the lung and over the opposite side.

144 DISEASES OF THE PLEURA

(c) Ægophony is heard at the upper limit of the fluid.

B. **Of pleurisy with scanty effusion.**

The physical signs vary in amount between a mere crepitating friction-rub without dulness and any or all of the signs of effusion. When contraction of adhesions sets in, there is collapse of one half of chest with

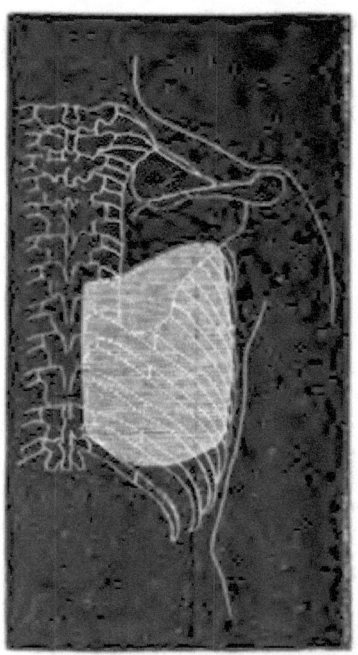

Fig. 24 A.—Diagram of the dulness in a case of right pleural effusion (from behind).

impaired movement and impaired vocal fremitus with great displacement of heart's apex-beat, patches of dulness, or general impairment of resonance. Tubular, bronchial, or sometimes amphoric breathing over the thickened pleura, and bronchophony; in fact all the signs pointing to consolidation may be present.

It is very difficult to diagnose between a chronically thickened pleura and basic consolidation of the lung.

Terminations.

   A. **In complete recovery,** with or without some adhesion.

This is the rule in dry pleurisy, and may occur also when the absorption of the exudation in other forms is rapid, although it is comparatively seldom that an acute pleurisy with effusion leaves the lung and chest in exactly the same condition in which it found them; if the effusion last beyond a few weeks complete recovery without shrinking is unlikely. The most complete recovery may be expected when the fluid has been carefully removed early in the course of the effusion.

   B. **In incomplete recovery.**

      a. In retraction and shrinking of the chest, which results from the lung being unable to expand properly in consequence of its having been too long or too firmly bound down. This termination, if the patient be otherwise healthy, may be looked upon as fairly favorable, as after a time the lung becomes partially expanded. As a result of its incomplete expansion, however, the chest wall falls in, the intercostal spaces are diminished in depth, the shoulder drops, and the spine may be curved with convexity to the sound side, the diaphragm is drawn up, and the heart's apex, when the pleurisy has been on the left side, may beat not far below the clavicle.

      A much better result than this may be expected when the fluid has been removed by operation, even if complete recovery does not occur.

      b. **In empyema.** We must also consider the possibilities of such an event. Supposing:

i. **The empyema has been tapped**; with careful after-treatment a very good result may be expected, the discharge becoming less and less and finally ceasing, the lung gradually expanding, and the external orifice closing up. The discharge often goes on for an indefinite time. A chronic discharge is more common, however, if

ii. **The empyema point externally**, the symptoms of which condition have been mentioned. If the position of the opening be in the third or fourth interspace it is unfavorable, as the discharge cannot then be perfectly free. But even if the empyema spontaneously discharge it is quite possible that, as in i, little shrinking of the chest wall will result.

iii. **The empyema may point through the lung** and be discharged through a bronchus. In such cases there may be previously symptoms of a slight pneumonia with rusty sputum, &c., or it may occur without any warning whatever, a copious discharge of pus taking place after a fit of coughing. Even in such a case there may be after a time recovery with more or less contraction of the chest; **pyopneumothorax** results, and sometimes death from suffocation follows:

iv. **The empyema may point through the diaphragm** into the peritoneum producing peritonitis, from which recovery is unusual, or into the pericardium (?), or some neighbouring organ.

c. In phthisis, either pneumonic, tubercular, or fibroid.

The relation between **phthisis and pleurisy** may be threefold:

1. The phthisis may be the cause of pleurisy by

the lung affection coming to the surface and producing a local irritation and so inflammation of the pleura; the phthisis being either pneumonic or tubercular; or by entrance into the pleura of irritating pus from a cavity after a small or large rupture.

2. **The phthisis may be the effect of pleurisy.** It very frequently happens that an attack of pleurisy is the antecedent of phthisis, and it is at once inferred that pleurisy causes it. This is not always the case, as the pneumonic or tubercular consolidation may have been slight and have caused the pleurisy. But in other cases phthisis appears to result from pleurisy, and then may be **tubercular**, probably from infection of the neighbouring lymphatics with the caseating material of the pleurisy, which in many cases, according to Niemeyer, may soften and break down. I believe, too, that **pneumonic** phthisis may result from the presence of the irritation of the thickened pleura. **Fibroid** phthisis is especially likely to follow from the compressed lung being the seat of a slow inflammatory process or **interstitial pneumonia**, or by the lymphatics being infected from the caseating material by the tubercle bacilli producing slowly-growing tubercles with development of fibroid tissue and few cells.

3. The **phthisis and pleurisy** may both result from a general pulmonary and pleural tuberculosis.

    d. In **amyloid disease** from the long-continued suppuration.

    c. **Death** may result in pleurisy from a variety of causes:

    i. In acute cases rarely from œdema of the lung and subsequent carbonic acid poisoning, or from excessive obstruction to the passage of blood through the lung by means of which the right heart becomes engorged with blood.

148 DISEASES OF THE PLEURA

ii. **From bursting of the empyema** into the peritoneum or neighbouring organs.
iii. **From** long-continued **hectic fever**, as in empyema if not relieved naturally or by operation.
iv. **From some variety of phthisis.**

Diagnosis.

I. Pleurisy *versus* Pneumonia.

Rigors and other symptoms of fever.

Seldom excessively marked; temp. seldom high; no herpes.	Very marked; temp. often very high; herpes about lips.

Crisis.

Rare.	Common.

Stitch in the side.

Almost invariable.	Rare, unless pleurisy is present.

Cough and sputum.

Dry cough, or with frothy sputum.	Rusty sputum, with cough.

Congestion of liver with jaundice.

Seldom, if ever.	Common.

Physical signs.

Intercostal spaces.

Flattened or prominent.	Not prominent.

Apex-beat.

Displaced towards the sound side.	Not displaced.

Friction fremitus.

Often present.	Absent.

Vocal fremitus.

Absent or diminished.	Increased.

Dulness.

Upper limit may be a straight or curved line, and may alter in position (p. 143); the dulness is resisting and absolute, and sometimes extends beyond the midsternum.	Upper limit a curved line, following the direction of the fissures of the lung, not altering in position, not resisting and absolute, not extending beyond midsternum.

Auscultation.

Absence of breath-sounds, or distinct tubular breathing and ægophony, &c.	Characteristic crackling crepitation, bronchial breathing, bronchophony, and crepitatio **redux**.

II. Pleurisy v. hepatic enlargement.

Hepatic enlargement seldom encroaches much into the chest, and if it does the dulness is increased upwards, more in front than behind. If the liver be enlarged it can be felt moving up and down with respiration, but if the pleura be filled with fluid the diaphragm is practically fixed, and the liver moves very little. With an enlarged liver the lower ribs are everted, but the intercostal spaces are natural.

III. Similarly with **enlarged spleen** the area of dulness varies with respiration, and the spleen may usually be felt on abdominal palpation.

Prognosis.—As will be understood from the above description of the terminations of pleurisy, the prognosis depends chiefly upon the nature and extent of the affection, dry pleurisy being of course as a rule not at all dangerous to life or health. Of the varieties with effusion we may tabulate some of those elements in prognosis which we consider as respectively good or bad.

Good.	**Bad.**
When the affection is acute in a healthy adult, with a distinct crisis (rare).	1. Excessive effusion, with great pressure on heart and veins, with blueness.
Rapid decrease of effusion.	2. Purulent effusion, on account of its tendency to complications.
Childhood.	

3. Implication of the opposite side, either with pleurisy or with bronchitis.
4. Presence of persistent fever with chronic effusion, wasting, and general asthenia.

5. Cyanosis and dropsy from any cause.
6. When the pleurisy comes on in course of any other blood disease.
7. Complications, as bursting of an empyema.

It may be noted that pleural effusion sometimes causes sudden death from rapidity of effusion. For example, we recently saw a case of empyema in a man who, after much persuasion, had consented to be operated on with a needle only. A large quantity of pus was removed; in a couple of days the pleura had refilled, and the patient refused his consent to a second operation, however mild, and died quite suddenly in a day or two afterwards.

Treatment

1. **Of pleuritis sicca.**—In dry pleurisy, although the pain is often quite out of proportion to the gravity of the affection, the treatment is simple, and is to be directed to **relieving the pain** and **diminishing the movement of the chest.** These objects may be accomplished:

   (a) By any of the following applications:—a **blister, a mustard plaster,** or the application of **iodine** (Lin. Iodi) to the painful part will generally suffice to relieve the pain. **Leeches** to the affected side; **cold compresses and hot poultices,** each have their advocates.

   (b) By **strapping** the affected side, in order to prevent the costal movements of that side; or **bandaging** the chest in order to prevent costal inspiration altogether.

2. **Of pleurisy with effusion.**—In all of the varieties of pleurisy with effusion, much of the treatment has to depend upon symptoms.

   (a) **Pain** has to be relieved by one of the methods above indicated.

   (b) **Fever,** which is seldom severe in uncomplicated pleurisy, may be met by the exhibition of quinine or digitalis or other anti-pyretic. The question of bleeding for this purpose is too wide to be discussed here, but in certain cases where cyanosis, dropsy, or dyspnœa arises in connection therewith, it certainly is admissible to let blood from the arm; such conditions are, however, rare.

(c) **Cough** may be relieved by mild opiates, including Tinct. Camp. Co , and by ipecacuanha and squills.

(d) **The effusion**, however, is that which requires most attention, and the first efforts should be directed **towards** promoting **its absorption as** soon as the diagnosis has been rendered certain, **by the withdrawal of some of the fluid by means of a morphia syringe if necessary.** For this purpose we employ:

    Purgatives of the watery, hydrogogue class, *e.g.* jalap, scammony, acid tartrate of potassium, or what not.

    Diuretics, such as broom, acetate of potassium, or spirit of juniper.

    Alteratives, especially iodide of potassium (gr. v to x ter).

    Local application of Lin. Iodi, or in some cases of blisters.

    Sudorifics.

    Restriction of the fluids of the diet is believed to have a good effect by some.

If, in spite of the medicinal treatment, the fluid does not diminish, it must be removed from the chest by the operation of **tapping**, called **paracentesis thoracis** or, shortly, **thoracentesis**.

Indications for paracentesis thoracis.—The circumstances under which it is advisable to perform the operation of tapping the chest may be summarised as follows:

(a) All cases of purulent effusion should be tapped. As this is a sure indication for the operation (at any rate in idiopathic pleurisy) great care must be taken to watch a case of pleurisy, and particularly the morning and evening temperatures, in order that no evidence of the presence of pus be overlooked.

(b) When there is great dyspnœa in acute pleurisies in consequence of excessive effusion, either from its pressure on the sound lung, or by its pressure upon the heart. In the latter case, especially when the heart is much pushed over to the opposite side, considerable obstruction to the venous circulation accompanied by blueness of the lips and tongue occurs. The question as to when the dyspnœa is sufficiently great to justify operation is an open one, and different answers would be returned to it by authorities according to the degree of importance attached by them to the severity of the operation. As for myself, when a question of the kind arises, considering the great success of the operation in my own experience, I should advise that it should be done as soon as the breathing becomes distinctly embarrassed, especially if there is evidence of much dislocation of the heart.

(c) When the fluid is slow in undergoing absorption or when the patient's strength appears to be too severely taxed by the effort of absorption, the operation is indicated. As to this indication for tapping in chronic pleurisy, it must be understood that no rigid line can be drawn as to the time which must be allowed to elapse before the operation is done. Many advise that only one week's interval shall be given after the fluid has ceased to increase before removing it. Possibly two weeks' grace may be permitted unless any symptoms become urgent.

Are there any contra-indications to the operation? There are scarcely any absolute contra-indications. But it is generally unnecessary to tap if the effusion is small, supposing there are no urgent symptoms, and in boys from twelve to seventeen years of age, who have been exposed to vicissitudes of the weather and to rough living, and in whom, says a writer on the subject, pleural effusions are common, tapping is seldom necessary, even though the effusion is excessive, as they tend to recover on good and nutritious diet and tonics without operative interference. My experience confirms this. Is it necessary to tap in symptomatic pleurisies? Here, again, the rule must hold that if the pleural effusion be obviously very great, in fact so great as to be evidently " adding to the distress of a patient already very ill," paracentesis should be done. This rule applies to all secondary pleurisies.

Double pleurisy is a condition in which one cannot lay down an absolutely rigid rule. I was once present at a consultation about a case of double pleural effusion, when the distinguished surgeon who had been called in refused to perform the operation of paracentesis. He said he considered the operation unjustifiable under the circumstances of the case, and as very likely to produce immediate death. This, however, must not be taken as a precedent, as it is obvious that there are cases of double pleural effusion in which it would be wrong to hesitate about operating, but the entrance of air into the pleural cavity must in such cases be most carefully guarded against.

Having determined that an operation is necessary two questions then arise: 1. As to the position of puncture, and 2. The kind of operation to be done.

    1. **Position of puncture.** — Some difference of opinion exists as to the best place for the operation of paracentesis. The largest number of authorities recommend that the puncture should be made in or near the mid-axillary line in the sixth or seventh interspace, according as the operation is needed on the right or left side respectively. Laennec preferred to punc-

ture in the fifth interspace a little anterior to the digitations of the serratus magnus, and with him not a few agree at the present day.

Lastly, puncture in or near the eighth interspace postero-laterally is advocated by many. Porritt, for example, suggests as the most convenient spot the seventh or eighth interspace, about the junction of the anterior two thirds with the posterior third, as a part where the muscular covering is less thick than in other situations. My own experience had been almost entirely of the first position until quite lately, but the last situation for puncture has recently been adopted at the Victoria Park Hospital with good success.

2. **Varieties of the operation.**—The operations of paracentesis thoracis may be divided into two classes:

    a. Those in which the opening into the chest is not free for the air to enter.

    b. Those in which the opening into the chest is free, and no precaution is taken to prevent the entrance of air.

The former variety of operation must be undertaken in every case in which the purulent nature of the enclosed fluid is doubtful, the latter in cases of empyema only. Knowing that the air contains an enormous number of saprogenous micro-organisms, it is not wonderful that when it is allowed to pass into the pleura in cases of serous effusion the effusion for the future becomes purulent, obviously a disaster to be guarded against by all possible means. In purulent effusions it is always better to incise the pleura freely in order that the pus may be allowed to drain away as soon as it re-forms.

Let us now consider the varieties of the operation in turn:

(*a*) **Paracentesis thoracis with closure of the opening.**—Always to be tried first of all, if the kind of fluid in the chest is a matter of even the slightest doubt, as well as of course in cases of undoubted non-purulent effusion.

The operation is performed as follows:—The patient is placed in a semi-prone position, near to the edge of the bed corresponding to the side upon which the operation is to be done. His head is supported by pillows, or he may lie down, inclining to the sound side. He should then be told to raise his arm above

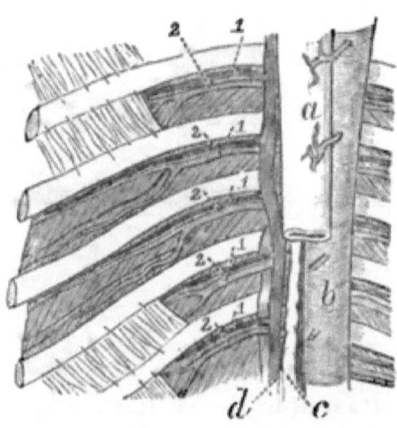

FIG. 25.—Posterior part of several intercostal spaces, to show the relations of the intercostal arteries within them. *a*, Œsophagus with arterial twigs from aorta; *b*, ramifying over it; *c*, thoracic duct; *d*, vena azygos; 1, intercostal vein; 2, intercostal artery. From the front.

his head, and the point where the puncture is to be made should be selected and marked. It must be in the middle of the interspace to avoid the intercostal arteries (Fig. 25). It is unnecessary that the patient should be under the influence of an anæsthetic, and indeed unadvisable.

The opening is to be made with a medium-sized trocar and cannula (Figs. 26 and 27), which is connected with an india-rubber tube two or three feet long, the free end of which is placed in a vessel at a convenient distance from the bedside beneath the surface of some water or carbolic lotion. The trocar can either be made in connection with the cannula, so that when it is withdrawn for a certain distance through an air-tight collar the fluid is able to run

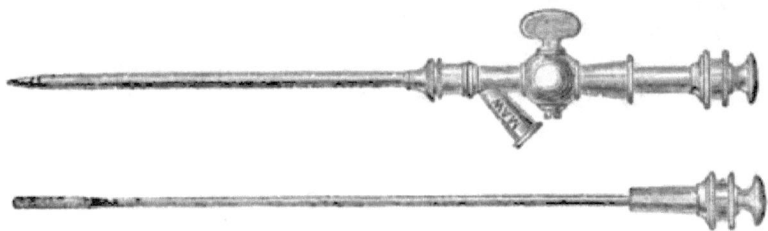

FIG. 26.—Potain's syphon-trocar and cannula.

FIG. 27.—Trocar and cannula. The former can be withdrawn by pulling out the milled head to the right, which is attached to the stout stilette, working in an air-tight collar, carrying the trocar.

through the trocar and through the india-rubber tube into the receiving vessel (Fig. 28), or it can be thrust through the india-rubber tubing into the cannula, and withdrawn in the same way, the hole in india-rubber made by its passage being, by means of the elasticity of the rubber, obliterated; or, thirdly, the cannula, by being sharpened at its extremity, is its own trocar (Fig. 28). By any of these means the fluid readily flows from the chest at first because it is under pressure, and afterwards owing to the syphon action of the trocar and tubing. Generally speaking, however, in my experience it is better to connect the cannula with an **aspirator** bottle (Fig. 28), which can be exhausted with an air-pump syringe and fitted with an arrangement of stopcocks, so that the bottle

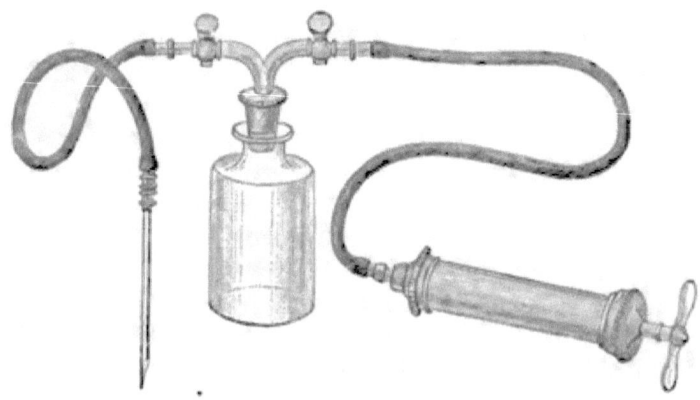

Fig. 28.—Aspirator. With exhaustion-syringe attached. The stopper is made of india rubber.

may be emptied of fluid when nearly full and re-exhausted.

As a preliminary to the introduction of the trocar, two things may be done if considered requisite:

(1) The **place of insertion may be locally anæsthetised**, either by the ether spray apparatus, or by means of a piece of ice dipped in salt held firmly against the skin, as recommended by Dr. Douglas Powell, or by solution of cocaine painted over the spot.

(2) **A small incision** through the **skin** may be made with a sharp scalpel. This incision is believed by some to be of considerable service, but by others as both unnecessary and even likely to lead to the entrance of air into the chest.

In the cases of non-purulent effusion it is not necessary to remove the whole of the fluid in the chest, indeed to do so is considered very unwise.

When sufficient fluid then has been withdrawn the skin is pinched up around the cannula during the process of removal, and a piece of thickly-spread, warmed plaster or a piece of carbolized lint soaked in collodium, is immediately placed over the wound, in order that no air may enter, a layer of cotton wool is then adjusted, and the side securely bandaged. Instead of the bandaging, firmly strapping the affected side may be employed.

The after-treatment consists simply in keeping the patient quiet in bed, of the administration of tonics, alteratives, and the careful watching in order to detect recurrence of the exudation.

Several other methods of tapping the pleura without allowing the entrance of air, have been recommended. One, suggested by Dr. Philip Hensley ('St. Bartholomew's Hospital Reports,' vol. xvii), is more intended for the treatment of empyema than for that of non-purulent effusion, but may be mentioned here.

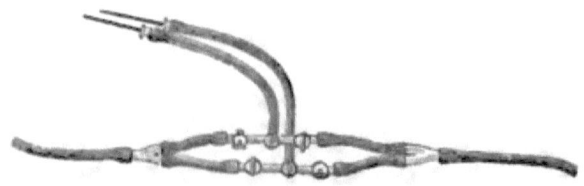

Fig. 29.—Hensley's apparatus for tapping and washing out the chest. The tube to the left is continued upwards to a vessel containing water some feet above the bed; the tube to the right is continued into a vessel containing water or the antiseptic solution on the floor. The tubes passing upwards are connected to needles which are pushed into two contiguous intercostal spaces. The stopcocks are so arranged that the fluid may pass in or out of the chest at will.

Two fine cannulæ, each having a connection with a receiving bottle, in which is some antiseptic solution, and with a reservoir of carbolic or some other antiseptic solution, raised three feet above the level of the bed. By means of taps, the direction of the current in each cannula (Fig. 29) can be at any moment reversed so as to be either outwards or inwards, and in this way the blocking of the tubes, which is not infrequently a cause of want of success in paracentesis thoracis, may be obviated. In order to thoroughly wash out the pleural cavity many pints of antiseptic fluid may have to be used (perhaps nine or ten).

Precautions to be observed

1. Perfect antiseptic methods ought to be followed, and the instruments should be chemically clean. Some urge that the carbolic spray should be always in action during the operation.
2. Too rapid or complete removal of the fluid should be avoided, so that when an aspirator, Potain's or another, is used the pressure should be carefully regulated.
3. If the cannula become stopped, the careful passage through it of a metal wire, or of the trocar may possibly remove the obstruction, or possibly increasing the suction power of the aspirator may have a similar effect.
4. In order to minimize the danger of syncope, the administration of a stimulant—a little wine or brandy—is recommended.

Complications.—It is satisfactory to be able to assert that in the majority of cases of paracentesis there are no complications worthy of mention, so that the operation of tapping the chest may be undertaken with confidence by one who pays the necessary attention to the details. Some of the complications which concern the physician as well as the surgeon are collected by Porritt and should be noted.

  i. **Syncope.**—This complication occasionally arises, but from what cause is uncertain, possibly from mere fear or nervousness or a tendency towards fainting, or from the pressure upon the vessels of a large surface being suddenly relieved, or from irritation of the pleural nerves, reflexly causing the heart to stop.

If the patient show signs of faintness the operation must be abandoned for the time.

ii. **Hæmorrhage.**—Blood may come either from rupture of the vessels in the false membrane or by puncture of the expanding lung, or possibly, but very rarely, from puncture of the intercostal arteries.

iii. **Cough and difficulty in breathing** are not infrequent, probably from the entrance of more blood into the pulmonary tissue after the pressure is relieved. These symptoms may be treated with success, at one time by change in the position of the patient, at another by a morphia injection, by opium, or by an ether draught.

iv. **Sudden death** from shock or similar condition sometimes happens. A few cases of this kind have been recorded by Watson and others.

v. **Wounding of neighbouring organs**, *e. g.* the liver, even if it be done, appears to do little harm, supposing the trocar be a fine one. If the heart be wounded, however, the consequences might be severe.

vi. **Surgical emphysema** is possible but very unlikely.

vii. **Albuminous expectoration and œdema of the lungs** very rare.

viii. **Hæmaturia** also rare.

ix. **Pyrexia** very possible but rare. Porritt says, "My experience has been that paracentesis relieves fever, and the symptoms called febrile, and does not induce them when absent."

(*b*) **Paracentesis thoracis with free opening in purulent effusions.**—When the removal of the whole of the fluid is advisable and necessary, and the hope that its re-accumulation will not occur is slight, aspiration of the chest is contra-indicated. It becomes necessary to make a free opening in the chest wall in order to evacuate the pus. This should be done with a sharp scalpel, and, according to most authorities, somewhat lower down and more posterior than in the former operation of aspiration, *i. e.* in the seventh or eighth space in the posterior axillary line. The incision should be made obliquely, and should be about an inch long. For convenience' sake, it may be as well to insert a cannula through the wound in order to carry off the pus. It is best to do the operation under antiseptic precautions. When this is the case, after the pus has been removed as completely as possible, a drainage-tube is inserted and fixed to the side by silk bound down with pieces of strapping about three or four inches long, the wound is covered with antiseptic gauze, and the chest is bandaged with antiseptic bandages. After a short time the absorbent gauze will have become saturated with pus and redressing will be required, but if antiseptic precautions have been properly attended to it is claimed for this method, that after the first dressing one full day may elapse before the next, and subsequently two or three days, the drainage-tube being washed on the third dressing, and on all subsequent occasions. The latter has to be shortened as the lung expands, and as a rule, finally drops out. According to our own experience, washing the chest out daily with a weak solution of carbolic acid 1 in 80, or with some other disinfectant is to be strongly recommended. The washing should be done by means of a raised pressure bottle, and the method of Dr Hensley appears to promise considerable success in suitable cases.

In many cases of empyema it becomes necessary to make a second or counter-opening in the chest in the interspace either above the primary incision, if it is low down, or below, if it is high up, in order more conveniently to wash out the chest. Some surgeons with much experience of chest surgery advocate that it should be done at the first operation. If this plan be adopted, a drainage-tube of india rubber is passed through both openings, and the ends are tied together outside the chest. A favourite dressing in such a case is carbolised tow.

The **after-treatment of cases of** pleurisy with effusion, both those in which paracentesis has been performed, and those in which no operation has been deemed necessary, should be directed to improving or keeping up the general health of the patient by good food, tonics, cod-liver oil, change of air and the like, and to preventing permanent deformity of the chest. This latter object may be attained to a considerable extent by directing that the muscles, especially those of the arms attached to the chest, should be regularly, patiently, and judiciously developed by gymnastic exercises, rubbing and massage, also that the patient should be encouraged in the exercise of increasing his respiratory capacity by taking a definite number of deep and regular inspirations at certain times daily.

II. Hydrothorax

A condition in which there is an effusion into the pleura of thin serous fluid not produced by inflammation. It is apt to arise in the course of general dropsy. The affections which are most likely to be complicated with this effusion are *chronic or acute diseases of the lungs*, which produce obstruction on the right side of the heart and general venous engorgement; *diseases of the heart*, both valvular and muscular; *chronic disease of the kidney* with albuminuria; *malignant malarious disease and chronic dysentery*. The fluid varies in amount but occurs in both pleuræ; it is generally movable but may be encapsuled. The pleural structure itself is opaque, white, and swollen, as is also the subserous tissue.

Symptoms.—When with the ordinary symptoms and appearance of general dropsy considerable *dyspnœa*, greatly increased on exertion, appears, it becomes necessary to carefully examine the chest;

and if the *physical signs are those of double pleurisy* a diagnosis of hydrothorax may be arrived at. As a rule, however, the intercostal furrows are not obliterated, neither is the heart displaced.

Treatment.—The treatment should be directed to the main disease, but if the fluid is excessive it may become necessary carefully to perform paracentesis.

III. Diaphragmatic Pleurisy

When inflammation attacks the diaphragmatic pleura it gives rise not unfrequently to such symptoms as mark it out from costal and pulmonary pleurisy and justify the consideration of it apart from them. Laennec has classed this form of the affection under the head of partial pleurisies, which he said might occur at any part of the surface of the lungs but which were found most frequently in three situations, viz. (1) in the fissures between the different lobes; (2) in the space between the base of the lungs and diaphragm; and (3) at the posterior-inferior and lateral part of the cavity of the pleura. In many cases the effusion is found to be puriform, but, as one would imagine, the inflammation may vary from the slightest roughening of the pleural surface to adhesions or localised collections of pus. When slight the affection is not uncommon. It is most frequent in women of adult age and is generally unilateral. According to Dr Donaldson,[*] jun., who has gone carefully into the etiology of the affection, it may be caused by or may accompany *pneumonia, pleurisy proper, hepatitis and peritonitis, wounds of the chest and abdomen, fracture of the lower ribs, blows at the base of the chest, currents of cold air on the perspiring body, immoderate laughter, continued crying and weeping, the suppression of rheumatism, the healing of old sores, tight lacing, and corsets.*

[*] "Diaphragmatic Pleurisy," 'Amer. Journ. of Medical Science,' April, 1886.

The secondary inflammation of the substance of the diaphragm is not uncommon.

Symptoms.—These may be at the beginning as in ordinary pleurisy.

- **Shiverings and fever** of more or less severity, but the following are the most characteristic symptoms.
- **Pain.**—More or less acute, sometimes of a very intense character, increased on movement, especially on cough, extending over the whole of the hypochondrium, radiating in all directions upwards to the neck and down to the flanks; tenderness on pressure, especially over the epigastrium, in the last **intercostal space behind**, along the course of the phrenic nerve, and at a spot, "*the diaphragmatic button*," one or two fingers' breadth from the middle line on a level with the **tenth rib**. The **pain and** tenderness are supposed to be due to the implication of the phrenic nerve in the inflammation, and through it by the transference of the sensation to the very **numerous** branches. These branches, it should be remembered, are distributed to the nerves of the neck, to the pericardium and to the pleura; it is connected with the sympathetic near the chest and with the fifth and sixth cervical nerves, the **solar** and hepatic plexuses, the peritoneum, and **adrenals**.
- **Orthopnœa and inclination of the body forward.**
- **Costal respiration**, hurried and jerky, the diaphragm being immovable.
- **Great anxiety of countenance,** with sudden change of features.
- **Cough.**—Frequent, dry, and very painful.

Of the more uncommon and less diagnostic symptoms may be mentioned:

Delirium. Pain in swallowing.
Hiccough. Risus sardonicus.

Nausea. Vomiting. Jaundice.

Course and termination.—The affection when primary generally ends in **recovery or** adhesion; when **secondary** it is of graver import. Sometimes the effusion between the base of the lung and the diaphragm is enclosed by the borders of the lung, already adherent, and a circumscribed abscess or empyema is thus formed. Such empyemata have been known to ulcerate through the lung and to discharge through a bronchus or through the diaphragm, and so to cause death.

IV. Pneumothorax

A condition in which the pleura contains air, almost always in conjunction with fluid, either purulent [pyopneumothorax] or non-purulent [**hydropneumothorax**].

Conditions of origin.—The gases which are found in **the chest cavity in a** large majority of cases, if not in every case, are derived from the external air. The air enters the pleura either through some perforation in the lung or through an opening in the thoracic walls; it is also possible for it **to enter** through a communication between the œsophagus **and pleura.**

The cases may be divided into:

i. **Spontaneous.**—Comprising all those in which the air has entered through the lung, in consequence of—

    (*a*) **Rupture or perforation of a cavity** in the lung, in advanced phthisis, or in less advanced **cases** from erosion of a superficial **consolidated** patch in the lungs. In rare **cases** superficial erosion of the lungs in gangrenous tumours **or infarcts.**

    (*b*) **Rupture of an emphysematous lung.**

    (*c*) **Erosion of the lung by discharge of an empyema through it,** as well as

    (*d*) **Pointing and discharge of an empyema** through the thoracic walls.

ii. **Non-spontaneous or traumatic:**
 (a) **By the operation of paracentesis thoracis** performed for any cause in which air enters the pleura.
 (b) **Wounds of the pleura, gunshot and otherwise,** including broken ribs, which injure the lungs, &c.

The above list does not include certain rare cases which are as much pathological as clinical curiosities, viz. (1) Pneumothorax due to openings in connection with the bronchus, œsophagus, stomach, or any other gas-containing cavity, and pleura from any cause e. g. growths or abscess. (2) Rupture of the lung in whooping-cough or in artificial respiration; neither does it mention the origin of the gases from decomposition of the fluid within the pleura, which was formerly believed by Laennec among others to be one of the most frequent sources.

From the statistics of my colleague, Dr Samuel West,[*] it would appear (1) that phthisis in one form or another accounts for no less than 90 per cent. of all cases of pneumothorax (excluding, I suppose, operative cases), and (2) that pneumothorax is present in 5 per cent. of all those who die of phthisis.

Pneumothorax may occur in phthisis, both early as well as late; "even at the commencement of the disease a small subpleural tubercle may soften and break through the pleura."[†] The commonest age is between twenty-five and thirty-five. As regards sex, it is common, but less fatal, in men than in women. It occurs rather more frequently on the left than on the right side.

In the statistics above referred to, drawn from the records of the Victoria Park Hospital for Diseases of the Chest, between the years 1856–83 inclusive, other points of interest are made out, for example

---

[*] "A Contribution to the Pathology of Pneumothorax," by S. West, M.D., 'Lancet,' vol. i, 1884, p. 791.
[†] Dr Powell, loc. cit., p. 131.

the **position of the perforation** in the fatal cases which were examined post mortem. When single, it was found ten times in the upper lobe, four times in the middle, and five times in the lower; in the upper lobe chiefly in the mid-lateral region. The **size of** the perforation generally was small and circular, one, two, or three lines in diameter, but one irregular opening was of the size of a five-shilling piece.

The **kind of phthisis** most often associated with it was the acute and rapidly advancing form. **Effusion** was present in half the cases, serum in four cases, sero-pus in eight, and pus in eight cases, the amount varying from an ounce or two to a couple of pints.

It should be remembered that in a general way there is not in pneumothorax any extra pressure when the opening is, as in the above cases, direct. This was proved by Dr Powell, who connected a manometer with the chest in some of such cases, and found no rise of the fluid in the tube. If, however, the **opening is oblique**, the air may enter the pleura during expiration, but the opening may be occluded during inspiration, and so it tends to accumulate in the pleura. In such cases the side becomes much distended and drum-like, and there is increasing rather than diminishing displacement of organs, as in Fig. 30, a case recorded by me in 'St Bartholomew's Hospital Reports,' 1887.

A pneumothorax is usually **complete**, except, perhaps, just at the apex where adhesions have formed, but it may be **partial** or **circumscribed**, when perforation of the lung takes place into a part of the pleural **cavity shut** off by adhesions from the remainder, or when the pleural cavity has been obliterated elsewhere by adhesions.

As regards the **gases** which have been found in the pleura if the opening is external, they are nitrogen, oxygen, and carbonic acid, in the ordinary proportion of the air. If the opening is internal, the gases are in the proportion of the alveolar air at first, but the

10 per cent. of oxygen quickly diminishes, and all but disappears, if the opening is not patent, and after a time the proportion has been found to be $CO_2 = 11\cdot16$, $N = 88\cdot35$, $O = \cdot49$, in one analysis. Sulphuretted hydrogen has been found in a case of a foetid pyopneumothorax.

Symptoms and Course.—The symptoms of pneumothorax may or may not be marked; if the latter, the pneumothorax is **latent**. Generally the attack is sudden, and the patient is seized with:—

Pain in the affected side, sharp and severe, in the region of the lower ribs.

Dyspnœa of a very marked kind, the **respirations** being rapid and shallow; this comes on very suddenly. The patient cannot lie on the affected side, but is compelled to sit upright in bed. The dyspnœa is due to the sudden cessation of breathing in the lung of the affected side, but also arises because at the same time the lung of the sound side is hyperæmic. Cough may occur, with sputum, more or less blood-stained.

Collapse for a time is frequently present; the patient is pale. There may be cyanosis and dropsy as well as other signs of engorgement of the right heart.

Pulse is small, soft, and frequent.

These symptoms are frequently followed by **fever of a subacute type** if the patient survive the immediate shock.

Terminations.—In death.

a. **Immediate**, from the shock. In the statistics above referred to 10 out of 39 died within the first day, 8 after a few hours, and 2 within the first hour, twenty and thirty minutes after the attack.

b. **Delayed.** Of the remaining 29, 19 died within the next fourteen days, and 6 in the second fortnight, *i.e.* 75 per cent. within a month.

In recovery. This is rare; when it occurs the aperture in the lung heals up and the gases in the pleura are gradually absorbed, and as this happens the lung expands. Recovery, at any rate partial, takes place most often in cases of circumscribed pneumothorax.

It should be recollected that pneumothorax is attended with few special symptoms, and is only to be made out on physical examination of the chest.

Physical Signs.—Inspection.—The affected side is bulged out and dilated, with obliteration of intercostal furrows. There is absence of movement. The cardiac impulse is displaced to the sound side.

Palpation.—The heart's apex-beat is displaced to the sound side; the liver is pushed downwards; the vocal fremitus is diminished or obliterated.

Percussion.—The percussion resonance is tympanitic at the upper portion of the chest, and this may extend to the opposite border of sternum. Posteriorly at the base of the lung there is nearly always some amount of dulness. The heart's dulness is dislocated towards the sound side. On percussing with coins there is the **bell sound** or the **anvil sound**.

The posterior dulness changes with position of the patient.

Auscultation — There is absence of vesicular breathing, possibly distant tubular breathing or rarely **amphoric**; on shaking the patient the **succussion sound**.

Metallic tinkling.
Metallic whisper **and echo**.
Sometimes, not often, **pectoriloquy**.

The breath-sounds may be heard over the apex of the lung if the pneumothorax is not complete there in consequence of the adhesion of the lung to the chest wall.

Diagnosis.—*a*. **From large superficial vomicæ** (in which we may get metallic sounds and amphoric resonance).

Differential diagnosis of—

	Emphysema (hypertrophous)	Pneumothorax	Pyo- or hydro-pneumothorax.
Inspection.	Both sides generally affected, chest fixed in state of more or less complete inspiration.	One side only affected, affected side fixed in state of more or less complete inspiration.	Ditto.
Form.	Marked prominence of angulus Ludovici. Intercostal spaces not prominent, frequently depressed during inspiration. Attachment of diaphragm mapped out by small cutaneous veins.	No marked prominence of sternum, but generally bulging in lower lateral region. Intercostal spaces prominent.	Bulging of lower lateral region very great. Intercostal spaces very prominent.
Movement.	Inspiratory movement short and slight, consisting rather of an upward movement of chest walls than of expansion outwards.	Greatly diminished on affected side. N.B.—The form and movement depend on the greater or less freedom of communication between the pleural cavity and bronchi. If this be great there is at first only slight deviation from health, but subsequently the affected side generally falls in. If communication be valvular, impeding exit of air, great distension results. Increased if side is distended or collapsed.	Ditto. Greatly increased at base of affected side.
Palpation. Resistance.	Diminished.	Depend on freedom of communication. If this be great these may be natural or even increased. Even if none they are rarely quite absent, being conducted along chest wall from opposite side.	Ditto.
Vocal vibrations.	Ditto.		Except that they are almost always absent at base.

DIAGNOSIS OF PNEUMOTHORAX 171

Position of heart.	Displaced down into epigastrium. Displacement often made more evident by hypertrophy and dilatation of right ventricle.	Displaced more or less towards opposite side; but if chest be collapsed it may be drawn over to affected side.	Ditto.
of liver.	Displaced down.	If right side be affected, displaced down if there be distension of the side.	Ditto.
Percussion.	Tympanitic or sub-. All over thickest with absence of cardiac or hepatic dulness.	Hyper-resonant throughout affected side.	Hyper-resonance above, sharply defined dulness below, horizontal, in erect position, but varying.
Ditto with coins.	Natural.	Bell sound.	Bell sound.
Auscultation. Respiratory murmur.	Inspiration feeble, expiration prolonged; the latter wheezing if cirrhosis coexist; if bronchiectasis, here and there cavernous.	Depends upon freedom of communication. I. May be absent or normal, or more or less amphoric.	Ditto.
Voice.	Feeble and muffled.	II. May be natural, absent, pectoriloquous, bronchophonic, or, in interscapular region, ægophonic.	Ditto at bases; absent or perhaps ægophonic.
Crepitation.	Absent.	Absent.	If opening be below level of fluid, gurgling.
Friction.	Often present.	Ditto.	Absent.
Special signs.	...	...	Succussion and metallic tinkling.

DISEASES OF THE PLEURA

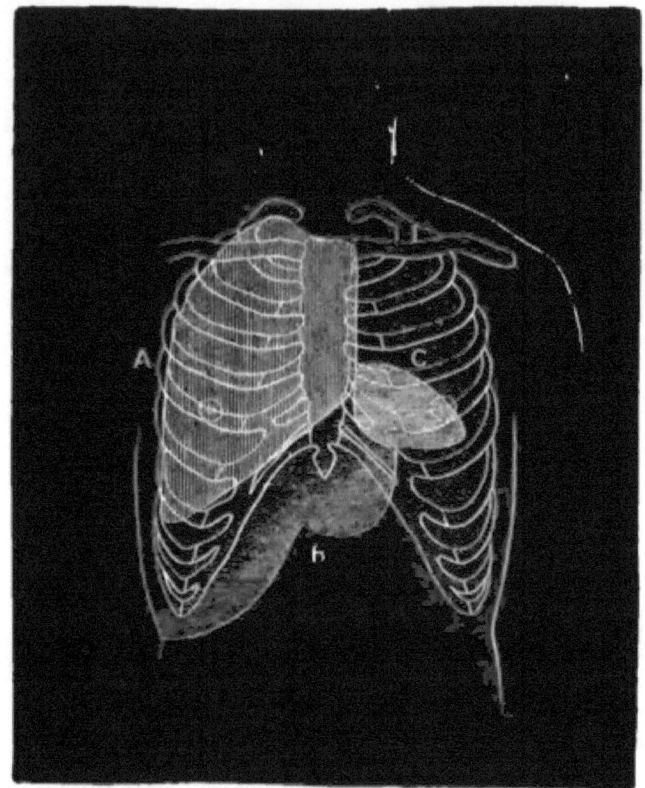

FIG. 30.—Diagram of the physical signs of the chest of Fanny P—. Pneumothorax of the right side.— A, The shading with longitudinal lines indicates the area of hyper-resonance, extending beyond the left border of the sternum; B, the shading indicating the displacement of the liver downwards; C, the transverse shading indicating the position of the cardiac dulness.

    i. The chest wall over a cavity is usually depressed.
    ii. The vocal fremitus is diminished.
   iii. There are nearly always numerous râles.
   iv. The neighbouring organs are not displaced.
    v. The pitch of the percussion note is altered by opening or shutting the mouth.

Treatment.—The treatment is chiefly symptomatic. When the patient is first attacked the shock is often so great that stimulants, either alcohol in the form of brandy or ether, are necessary, and it is then advised to give some opiate such as morphia subcutaneously.

Pain is relieved by poultices, leeches, or the application of cold.

Dyspnœa, if it become urgent, is relieved by tapping the pleura with a fine trocar and letting out the air. Venæsection is recommended by some, we think upon very doubtful grounds.

If the patient survive the shock, absolute quiet in bed in a well-ventilated, warm room; poultices to the side, or a plaster of belladonna or strapping; nourishing light food, such as milk, beef-tea, chicken broth, given in small quantities very frequently, should be enjoined; and a moderate amount of stimulants should be given. It is nearly always necessary to continue for some time to exhibit opium in some form. If the fluid in the pleura increase it may become necessary to remove it by paracentesis.

V. Hæmothorax

The presence of blood within the pleural cavity, may be suspected if considerable dulness on percussion and dyspnœa arise soon after the fracture of one or more ribs, or as a wound of the chest. The blood may come from the intercostal artery, or from the tissues of the lung itself. The subject is treated of in hand-books on surgery.

VI. Carcinoma and Tubercle of the Pleura

are treated of under the heads of the same affections of the lung.

CHAPTER VI

DISEASES OF THE RESPIRATORY MUCOUS MEMBRANE

As it is practically impossible to separate affections of the trachea from those of the larger bronchi for purposes of description, in this chapter will be included the diseases of the trachea and bronchi.

Catarrhal Inflammation. — [Tracheitis and bronchitis; catarrh of the trachea and bronchi.]

Causes.—There are certain conditions which are said to predispose to catarrh of mucous membrane; they are as follows:—(1) Malnutrition. (2) Previous attacks. (3) Too great care to prevent catching cold, by wrapping up too much or remaining indoors, &c., appears to create a delicacy of constitution which predisposes. (4) Idiosyncrasy, perhaps due to the possession of a thin skin or to a special tendency to sweat.

The exciting causes are (1) Wet or cold. The mere change of weather from warm to cold or from dry to wet at once very largely increases the number of catarrhs. They may arise from chilling of the surface or from breathing in cold air. In these cases the catarrh almost always but not invariably begins as a cold in the nose and spreads down the mucous membrane to the trachea and bronchi. (2) Inhalation of irritant substances, such as dust, medicinal substances, *e. g.* ipecacuanha. (3) Epidemic influences, when the bronchial catarrh is merely a part of an attack of influenza. (4) Other diseases. It often occurs in the course of other acute or chronic affections, *e. g.* measles, typhoid fever, and the like. Bronchial catarrh is also common in gout and also in chronic Bright's disease (chronic interstitial nephritis). (5) **Local** irritation, as the presence of morbid growths in the neighbourhood, *i. e.* **of cancer**; of tubercles; **of ulceration**; or of pressure, as of an **aneurysm.** (6) Congestion of the bronchial vessels, either active or passive, especially in heart disease.

Pathology and morbid anatomy.—The process of catarrhal inflammation consists of an hyperæmia of the mucous membrane, and of inflammatory swelling of the tissue from an exudation containing few **leucocytes**, into it. Combined with this there is **excessive secretion from the** surface, consisting chiefly of **immature cells** and, possibly, a few colourless corpuscles, with an extra mucous secretion from the bronchial mucous glands. Often it appears that the inflammatory element **scarcely enters at all**, but there is hyperæmia simply, and excessive mucous secretion.

The post-mortem appearances in death from—

A. **Acute catarrh** are not very marked. The mucous membrane is mottled or red, clouded, opaque, relaxed, and **tears** easily. The bronchi are diminished in calibre. The surface of the mucous membrane is dry or covered with scanty, tenacious, transparent secretion, containing young cells and some columnar ciliated cells, but later covered with yellowish secretion containing immature cells. The lungs do not collapse unless the chest is opened.

B. **In chronic catarrh** there is more marked **colouration of the mucous** membrane from the **permanent dilatation of the blood-vessels**. The whole of the mucous membrane **is increased and** less easily **torn.** Microscopically it is **found that** the columnar **ciliated cells** are absent from the surface, their place being taken by immature **cells of** different shapes, never columnar. The basement membrane is swollen, and the tissue beneath is packed with **small inflam-**

matory corpuscles. These do not appear to be capable of passing through the basement membrane on to the surface.

In many cases the **bronchi are markedly dilated and** gape widely on section.

The **surface is covered** with a **yellow purulent secretion** or with a glairy tenacious mucus.

Ulceration, both diffuse and follicular, may be here and there distinctly made out in the latter case from the implication of the lymphoid tissue in the inflammation of the mucous membrane.

The affection differs remarkably in severity, according as it is **acute** or **chronic.** It is necessary to distinguish between the symptoms and course of the two conditions.

I. Acute catarrh

A. **When the trachea and large bronchi only are implicated.**

Symptoms.—The affection is usually, but not always, accompanied by—

Catarrhal fever.—Shown by shiverings or a single rigor, feelings of burning heat, headache, soreness of the limbs, pain in the joints increased on pressure, anorexia, furred tongue, constipation, quick **pulse, raised temperature,** 100°—101° F.

Feeling of constriction across the chest.

Cough short and wheezy.

Sputum at first scanty and difficult to bring up, then glairy and pigmented (sputum crudum), afterwards muco-purulent (sputum coctum) usually frothy.

Difficulty in breathing more or less marked.

On Physical Examination the chest movements are not naturally **free, and sonorous rhonchus** is heard over the chest. When the secretion is free there are **large crepitations or mucous rales.**

Course.—Usually towards recovery after a varying time, but sometimes the catarrh spreads to the smaller tubes.

B. **When the** smaller or capillary **tubes are implicated** [Capillary bronchitis].

    Symptoms are much the same as in the former case, but are much more severe.

    Cough more violent and distressing, there are prolonged paroxysms, during which the patient gets red or bluish in the face, strains violently, or even vomits.

    Sputum is heavy and sinks in water, but may be kept up by the adhering frothy mucus from the larger tubes

    Difficulty of breathing becomes more marked.

On Physical Examination there is whistling, wheezing sibilus, and when the secretion is free small crepitation (or subcrepitant râles).

Terminations in —

(a) Recovery in a week or so.

(b) **Incomplete recovery** when the disease passes into chronic catarrh, or, what is more serious, catarrhal pneumonia (see p. 230).

(c) **Death** in the typhoid state—especially likely to occur in the old and feeble—delirium, and coma preceding death. The tongue becomes dry and brown, the pulse small and irregular, the skin sweats; a rattling is heard all over the chest both on inspiration and expiration.

c. **In young children** in whom capillary bronchitis is not uncommon it is a very serious affection indeed. It generally supervenes, often very quickly, upon catarrh of the larger tubes. The symptoms are that the difficulty of breathing becomes greater and greater. There is croupy cough and whistling inspiration, and restlessness, but the fever may or may not be marked.

As the disease advances the child becomes more and more livid, the lower part of the chest is seen to recede during each inspiration, and if the condition is not relieved the patient may become comatose and die.

The **Physical signs** are the same as in the case of adults.

D. In the newly born and in very young children there are often in addition the symptoms of collapse of certain parts of the lungs. This is due to the unexpelled mucus in some of the smaller tubes acting as valves which prevent the inlet of air, but at the same time permit of its exit. If such collapse of the air-vesicles supervenes the child becomes livid, with pinched nose, dull eyes, cooled skin, lowered temperature, and flabby muscles. Unless the offending mucus be expelled by vomiting or by other violent expiratory effort, death from asphyxia is liable to occur.

II. Chronic Catarrh

Symptoms.—It is to chronic catarrh of the bronchi that most winter coughs are due. The symptoms usually begin about October or November, and last until April or May.

Cough is very troublesome, and is often very paroxysmal, being set up on the slightest irritation, e. g. cold air, smoke, fog, or what not. During the paroxysm the patient becomes red or blue in the face, the neck swells, the jugulars dilate, the eyes appear starting out of their sockets, tears flow down the cheeks, and the nose runs. The attack frequently ends in vomiting, and in bringing up of more or less phlegm.

Difficulty in breathing, becoming worse year by year, varying, however, as the weather varies, occasionally so severe as to threaten life, accompanied by audible wheezing or bubbling. As the dyspnœa becomes chronic the muscles of inspiration, both ordinary and extraordinary, become much hypertrophied. After a variable time there are signs of:

Dilatation and hypertrophy of the right heart, and when the former exceeds the latter we have symptoms of pulmonary and systemic venous obstruction with—

Blueness of the lips and tongue, and general cyanosis and dropsy, and the other consequences of that condition.

Physical signs are the same as those of acute bronchial catarrh.

Course and Terminations.—Generally as above indicated the affection tends to increase, and when once a person has been attacked by it, complete immunity for the future is extremely unlikely. Death may arise from—

(*a*) Cardiac failure, or
(*b*) Some intercurrent malady, or
(*c*) Further pulmonary disorders, including phthisis.

Prognosis.—*a*. If the larger tubes only are affected, and the patient is strong, there is little or no danger.

b. If the smaller tubes are also implicated the danger is greater, especially if the patient is very young or past middle life. As to capillary bronchitis in children, it is always serious.

c. Indications that there is collapse of the lung or that the inflammation has extended into the alveoli, must also be looked upon with great anxiety.

Treatment.—The treatment of the affection varies much according to the severity and stage of the disease, as well as to its cause.

A. In acute catarrh of a mild kind, as when the inflammation does not extend beyond the larger tubes, the patient should be kept in a warm room without draughts, the temperature of which should be maintained at about 65° F. A kettle of water should be kept boiling near the fire and the steam should be carried well into the room through a long spout.

> **Inhalations** of steam or of various medicated vapours, of which conium (Vapor Conii) and benzoin (Vapor Benzoini) are most useful, should be employed.
>
> **Diaphoretics**, such as Dover's powder, or Spiritus Ætheris Nitrosi, may be ordered in the very early stages.
>
> A mild **purge** is often of considerable use.
>
> **Counter-irritation**, by means of a mustard plaster, or with Lin. Terebin. Aceticum.

B. **In acute catarrh of a severer kind**, not only must the patient be kept to his room, but it is also advisable to keep him in bed. In case of a child, bringing the cot opposite the fire and enclosing it with a screen is a practice frequently adopted.

> **Energetic counter-irritation** should be persevered in.
>
> **Bleeding** by means of leeches or wet cupping from the chest is sometimes indicated in adults.
>
> **Emetics** of ipecacuanha are especially useful in children.
>
> **Expectorants**, carbonate of ammonia, squills, paregoric, and small doses of opium must be used as the secretion becomes more profuse. But the greatest care is necessary in the administration of opium, and it should be given in small doses only.

C. **In chronic catarrh**, those who suffer year by year from winter cough which tends to get worse, should be advised, if their means permit, to seek a more genial winter climate, should be ordered to wear flannel next the skin all the year round, and should not be allowed to go out in foggy, cold, or damp weather.

The treatment of the catarrh when in progress should be by means of—
- **Counter-irritation,** *e.g.* **Lin. Terebin. Aceticum,** Lin. Crotonis, Lin. Iodi.
- **Stimulant expectorants,** *e.g.* carbonate of ammonia, senega, squills, &c., balsams of Peru, Tolu and copaiba.
- **Inhalations** of carbolic acid, creasote, or tincture of benzoin.
- **Tonics,** *e.g.* strychnine, quinine—the former being often **most** useful.
- **Antispasmodics,** if there be much spasm superadded, *e.g.* iodide and bromide of potassium.

When the **bronchi are dilated** the stimulant expectorants are especially indicated, and when the expectoration is not free an **emetic often acts like a charm.**

In **bronchorrhœa** the resins and balsams and inhalations are also specially indicated.

III. Croupous Inflammation [Spastic or croupous bronchitis]

is a rare affection, characterised by **the formation** upon the surface of the bronchial mucous membrane of a false membrane of fibrinous material, and so distinguished from catarrhal inflammation, the hyperæmia and swelling of the tissue being, it may be, little different in both varieties. As a rule it arises from the spreading of the inflammation from the larynx in diphtheria, but it may arise idiopathically in the smaller bronchi. In the latter case it is said to attack young people (chiefly males) at the time of puberty. The false membrane when brought up forms a more or less complete cast of the tubes.

Symptoms.—
- Sputum.—This is the only certain symptom. The sputum is made up of regularly formed casts of the tubes, slightly blood-stained.

If this is coughed up once, it may be a second time or oftener.

Cough, often very distressing until the cast be got rid of, accompanied by—

Dyspnœa, which may be very intense until the expectoration has been accomplished.

Hæmoptosis may precede, accompany, or follow the expectoration of the cast.

Prognosis.—Death seldom occurs, but in a recorded case it was due to blocking of the trachea with the false membrane. Complete recovery is also rare.

Treatment.—If dyspnœa is great, an emetic should be given. Iodide of potassium is recommended to relieve spasm, and the various drugs said to be capable of dissolving the fibrin may also be used.

IV. Bronchial Spasm [Asthma. Nervous or spasmodic asthma]

An affection in which there is sudden spasm of the unstriped muscular fibres of the bronchial tubes, preventing the free exit of air. The spasm is usually reflex, and the efferent impulses are conveyed by the branches of the vagus distributed to the bronchial muscles. A similar spasm is said to occur from direct pressure on the vagus trunk. In the cases of asthma proper, the irritation in all probability arises in the bronchial mucous membrane, so that both the sensory and motor impulses are conveyed by the vagus.

Causes.—Spasmodic contraction of the bronchial muscles producing dyspnœa occurs in various conditions, but is most typically manifested in the idiopathic affection, which is known as true asthma. For the purposes of description Salter divides all such cases therefore into—

(*a*) Idiopathic or primary—true asthma.
(*b*) Symptomatic or secondary.

A. As regards true asthma, it attacks **males** more often than females, is **hereditary**, begins usually after **ten years of age,** but **may** begin in infancy and continue throughout life. First attack **may** sometimes be traced to a bronchial attack or a whooping-cough, **but** usually it appears to arise without any such cause. When once it has occurred, it may **recur** with distinct **periodicity,** *e.g.* weekly, monthly, or yearly from different and very various exciting causes. **Exciting causes of an asthmatic attack.**
  1. **Peculiar localities,** varying **with the individual,** high or low, moist or **dry, town** or country, seaside or inland. London, however, is usually less likely to excite attacks than the country, a moist **atmosphere than a dry, a low-lying** place **than one of higher** elevation.
  2. **Inhalations of various kinds,** of dust, smoke, medicinal powders, *e.g.* ipecacuanha.
  3. **Certain odours,** exhalations from animals, especially **cats;** scents of flowers or of new-mown hay; **the smell of cooking,** roast mutton, for example.
  4. **Certain directions of the wind,** varying in individual **cases.**
  5. **Anxiety of mind or mental disturbance,** sexual excesses, constipation, and distension of the bowels.

B. Symptomatic asthma may arise in almost any disease in which the terminations of the vagus nerves **are** affected, as **well as in** those when either vagus trunk is pressed upon. It may occur in morbus cordis, emphysema, pressure of a tumour upon vagus, also in cases of persistent thymus or enlarged glands. Also from other reflex causes, *e.g.* uterine affections. It is one of the phenomena of gout, and of interstitial nephritis.

Post-mortem Appearances.—In true asthma there is no gross lesion to be discovered. In symptomatic asthma the remote cause of the attack may sometimes be made out.

Symptoms.—(*a*) **Of an attack.** There may or may not be premonitory symptoms, for some hours before, such as drowsiness, depression, stuffiness about the chest or wheezing, or itching under the chin.

The attack may begin at any time of the day or night, usually in **early morning** about 3 or 4 a.m. Should the patient be asleep when it comes on, he has terrible dreams, and wakes up with a sense of suffocation, being unable to draw in a deep breath. He throws off everything which constricts his chest at all, and very likely rushes to open the window, at which he remains with his elbows on the window ledge, however cold the weather may be. The efforts to breathe are painfully manifest, all of the extraordinary muscles of inspiration being pressed into the service.

The inspirations are **slow** but **not deep**, but the expirations are very prolonged, with wheezing, hissing, whistling, or purring, **audible on** inspiration and **on** expiration, both to the bystanders and to the patient himself. The chest is in a condition of semi-distension, even at the end of expiration, and is hyper-resonant on percussion, and little air leaves the lungs. Sibilus is heard throughout the chest on auscultation. The pulse is small, weak, and often irregular. After a time, varying from a quarter of an hour to several hours, the attack gradually or suddenly subsides, with or without eructations, yawning, or cough. The cough, at first dry, is soon accompanied by expectoration of a little mucus, and crepitations are heard in the chest, with or instead of sibilus.

(b) **Between the attacks.** The growth of the youth who is asthmatic is delayed, and puberty is reached very late. The asthmatic patient is usually extremely thin, with shrill voice, prominent eyes, sallow complexion, high shoulders, and deformed chest. His breath is almost always short, unless he lives in a locality which exactly suits him.

Prognosis.—Death seldom occurs in a paroxysm, but rupture of a vessel in the brain has been known to happen. Cessation of attacks also may occur; as a rule, however, a patient once subject to these attacks remains so to a greater or less degree. The cessation of an attack is probably due to the action of the carbonic acid gas in the blood relaxing the muscular fibres of the bronchioles through the respiratory vaso-motor centre.

Treatment.—During the attack remove any tight clothing which may interfere with respiration, see that the ventilation of the room is free, and that the air is warmed; help the patient into a position of greatest comfort, and administer such drugs as shall have been found most useful in the case. The following may be enumerated as among those most frequently employed:

Himrod's powder,* ⎫ burnt, so that the patient
   Nitre paper   ⎭ may inhale the smoke.

Strong tobacco, smoked till the patient is nauseated.

Stramonium leaves ⎫ smoked as cigarettes.
  (Datura tatula) ⎭

Emetics, such as tartarated antimony, or ipecacuanha.

Strong coffee.

Chloroform, ⎫ give great temporary relief.
   Ether   ⎭

Lobelia inflata, belladonna, hyoscyamus, or morphine.

* This is a patent medicine; it is undeniably of much use in asthma.

Counter-irritation in form of mustard leaves, &c.

Iodide of potassium, given regularly between the paroxysms, greatly reduces their frequency and severity. It is also especially useful in bronchitic affections, in which there is a tendency to symptomatic asthmatic attacks.

When asthma is symptomatic of other diseases the attention should be directed to removing the irritation which determines the attacks.

According to some authors there is a form of asthma in which there is spasm of the diaphragm and not of the bronchial muscles. Sibilus is then absent, and there is obstruction of the expiration and not of inspiration.

Symptoms.—In this rare condition, of which we have had no experience, there is marked contraction of the abdominal muscles. The urine and fæces pass involuntarily. Expiration is very prolonged. On percussion extreme depression of the diaphragm is made out. On auscultation the respiratory murmur is inaudible.

V. Bronchial Dilatation [Bronchiectasis]

There are several distinct forms of dilatation of the bronchi:

1. In which the tubes, most frequently of the second and third degree, undergo a general dilatation, but the smaller are more enlarged proportionately to their normal calibre than the larger. This is the effect of chronic bronchitis. The surrounding lung tissue is emphysematous.
2. In which the tubes present here and there, either in their course or termination, a more or less globular dilatation, more marked, perhaps, on one side than the other. This form occurs also in bronchitis, but in cases in which there is also associated collapse, consolidation, and the like.
3. Where the cavities are irregular and surrounded by dense fibrous material, resulting from chronic

interstitial pneumonia, which may or may not be a sequel of chronic bronchitis.

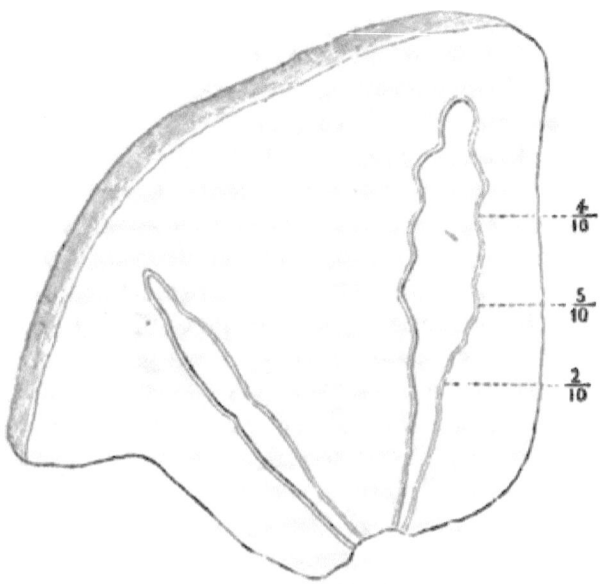

Fig. 31.—Diagram of a section of lung in which are two dilated bronchial tubes, the shape and relative size being strictly copied from a specimen in the Museum of St Bartholomew's Hospital.

4. Where the bronchiectasis is not, properly speaking, a dilatation of the bronchial tube itself, but is a cavity in the lung into which a bronchus leads, the lining membrane, however, being continuous from the tube into the cavity. The cavity is surrounded by dense fibrous tissue. In this form there may be several such bronchiectases in the lung or only one. They may be situated either in the apex or base.

Pathology.—A good many factors enter into the causation of bronchiectasis proper:

1. Dilatation of the tubes from accumulation of catarrhal products, preceded by a weakness due to
2. Atrophy of bronchial muscles from pressure of

the catarrhal inflammatory products beneath resistant basement membrane (Hamilton).
3. Forced expiratory efforts producing yielding of the walls already weakened.
4. Inspiratory efforts when a considerable portion of the lung tissue is impervious to air produces dilatation of the bronchi, if their resisting power is less than that of the lung.
5. Traction of fibrous cicatricial tissue surrounding bronchi, especially if the two surfaces of the pleura be adherent and thickened. Bands of fibrous tissue are then to be found dipping down into the substance of the lung, and joining the new fibrous tissue which surrounds the bronchi.

Generally bronchiectases are to be found more frequently in the middle or lower lobe of the lung than elsewhere. The sacculations vary in size from that of a pea to that of a walnut.

The inner surface of the dilated tube is at first smooth, the mucous membrane is thin, and the secretion may be glairy mucus only. Later on ulceration with roughening of the mucous membrane may occur, from which sometimes severe hæmorrhage and occasionally gangrene may result.

Symptoms.—When in the course of any of the affections likely to produce the condition, such as chronic bronchitis, pleurisy, or the like, the patient presents the following symptoms, as well as the physical signs to be presently mentioned, bronchiectasis may be diagnosed.

Sputum in large quantity, of a reddish-yellow colour, curdy, and of an *intensely foetid* odour, on microscopic examination showing immense numbers of putrefactive micro-organisms but no tubercle bacilli.

Cough, if not incessant, which is sometimes the case, may be of two kinds; for hours or even days the patient may cough very slightly, with slight frothy expectoration, but then may have a very severe spasm, in the course of which he will in a short time spit up a large quantity of the sputum above described. A paroxysm may often be brought on by the patient leaning his head over the side of the bed.

Clubbing of the fingers, when there is much accompanying fibroid induration.

Physical Signs.—If not obscured by the dilatation, being away from the surface of the lung or by surrounding emphysema. The physical signs are:

Inspection.—Flattening or contraction, with impaired movement.

Palpation.—The vocal fremitus may be increased or diminished.

Percussion.—There is dulness with increased resistance.

Auscultation.—There is feeble respiration, with moist râles on cough. Sometimes cavernous breathing and pectoriloquy.

Treatment.—The medicinal treatment of bronchial dilatation has been already indicated. Within the last year or so several cases of bronchiectasis of the lower lobe have been treated by incision and draining. The opening has been made in the eighth or ninth interspace posteriorly. Although in several of the cases, death from septicæmia took place, yet in the others the treatment may be said to have been highly beneficial. Dr C. Theodore Williams, who, up to March, 1886, had treated six cases by incision with three deaths, sums up the indications for the operation as follows:

1. In cases where antiseptic treatment of all kinds having failed to correct the fœtor of expectoration and to allay the harassing nature of the cough, and death by septic poisoning seems imminent.
2. Where all evidence goes to prove that the bronchiectases are confined to one lung, are situated in the lower lobe, and have overlying them an adherent pleura.

VI. Whooping-Cough

Etiology.—It is a specific disease, characterised by a catarrh of the bronchial mucous membrane and reflexly through the respiratory centres, by fits of paroxysmal cough, depending upon a peculiar hyperæsthesia of the mucous membrane.

It is **contagious**, directly and indirectly, is sometimes **epidemic**, occurs mostly in winter and spring, and is often associated with epidemics of measles and scarlet fever, is prone to attack children after two years of age; **one attack**, as a rule, **secures immunity**. Adults are very rarely attacked.

The specific poison is **contained in** the secretion and exhalations from the **diseased** membrane, and is believed to be a **special micro-organism** (micrococcus).

The exact relation of the poison to the disease is unknown. The inflammation of the bronchial mucous membrane is no doubt the local manifestation of the constitutional state. The bronchial catarrh is attended with marked hyperæsthesia, and the spasmodic closure of the glottis is brought about reflexly. The vagus fibres distributed to the mucous membrane of the bronchi are the conductors of the afferent, and the recurrent laryngeal nerves of the efferent, impulses.

Post-mortem Appearances.—Nothing striking is found beyond a certain amount of catarrh and the presence of the lesions which have caused death, not peculiar to whooping-cough.

Symptoms.—Three stages are usually described in this affection.

   I. Catarrhal stage, after a period of incubation lasting about ten days to a fortnight.

     Fever of somewhat severe kind, with sneezing, reddening of the conjunctiva, photophobia, running at the nose, dry lips and furred tongue, considerable constitutional disturbance, and an Irritating cough, occurring in paroxysms, at the end of which a large quantity of viscid, frothy mucus is expectorated.

II. **Whooping stage.**—After about ten days the catarrhal fever has subsided, and the characteristic whooping occurs with the cough.

There may be a **warning** of each coming paroxysm, such as—

Substernal uneasiness,
Tickling in the throat,

and then the **attack comes on.** The patient takes in an inspiration, the laryngeal muscles are contracted by violent reflex action, and the glottis is closed. The air is then driven out in puffs and jerks by the action of the expiratory muscles, bursting open the glottis by the violence of the impact of air against the cords. When the whole of the tidal air has been expelled, a prolonged crowing inspiration follows, causing the characteristic whoop.

Expirations and inspirations of similar character succeed the first several times, until a certain amount of viscid secretion is expectorated, with or without vomiting.

During the attack the straining is exceedingly violent, and in addition to extreme blueness of the surface of lips and tongue there may be—

Involuntary evacuation of urine or fæces.
Rupture and prolapsus ani.
Hæmorrhages, stigmata, &c., in the skin and mucous membranes, especially in the conjunctiva.
Rupture of membrana tympani.

The paroxysms vary in number in different cases, but are nearly always worse at night. About twenty to forty attacks in the twenty-four hours would be about the usual number, although there may be more. In the intervals, except just after a paroxysm by which he may be exhausted for a time, the patient is fairly well. Underneath the tongue there is generally to be found some ulceration of the mucous membrane with (or without?) rupture of frænum linguæ.

III. **Stage of decline** cannot be said to begin at any special time, but the former stage passes gradually into it after a very variable number of weeks, say four or five, more or less. The attacks become less and less frequent and severe. The sputum becomes muco-purulent, and vomiting is less frequent.

Relapses are common.

Physical Signs are those of catarrh of the bronchi.

Terminations.—1. **In recovery** after six weeks to six months.

2. **In incomplete recovery**, leaving behind—
Ruptures either umbilical and inguinal.
Emphysema and bronchitis.
Collapse of portions of the lung.
Catarrhal pneumonia.
Phthisis.

3. **In death from the above sequelæ or from** meningitis or apoplexy.

Diagnosis cannot be made with certainty until the characteristic whoop appears with the cough.

Prognosis.—Whooping-cough is a serious affection in young children, and, speaking generally, is most fatal in the youngest and most debilitated children. In older children it is not often fatal.

Treatment.—As the old idea that every child must get whooping-cough at some time or other is given up, isolation of the sick from those who have not had the disease must be carefully carried out, and localities in which whooping-cough is rife should, if possible, be avoided.

It is best during whooping-cough to keep children within doors and in an equable temperature, warmly clad, and if the attack be very severe, in bed. As there are no indications from the disease, it is necessary to attend to the general health, to administer good and nutritious diet, and to treat symptoms.

- **To relieve the severity of the coughing.**—As regards medicine, a mixture containing Tinct. Belladonnæ (m_v to m_x) and Pot. Bromid. (gr. iij to gr. v) we think, is very often of service in reducing both the number and severity of the paroxysms. It cannot, of course, be expected to cure the disease. For the same purpose other drugs are believed in by various authorities, *e. g.* arsenic, iron, ipecacuanha, hydrocyanic acid (m_j), bromide of ammonium, hyoscyamus, &c.
- **Emetic mixtures** are sometimes necessary when the secretion is not expectorated freely, and those containing ipecacuanha in the form of Vin. Ipecac. are most frequently employed.
- **Counter-irritation** of the chest with stimulating liniments is nearly always of service. The local application of a solution of silver nitrate to the larynx has been advocated by some.
- **Inhalations** of benzoin or conium are useful if the children are old enough to use them.
- **In** the stage of decline, as the paroxysms diminish in severity and frequency, **tonics,** quinine, iron, cod-liver oil; and eggs, milk, and wine in addition, to the ordinary food are of good service.

Nothing finishes off an attack of whooping-cough so satisfactorily as a change of air to the seaside.

VII. Influenza or Epidemic Catarrh [Catarrhus e Contagio]

Influenza is an acute catarrh of the mucous membranes of the eyes, nose, and respiratory passages generally, chiefly remarkable for its sudden onset, its wide-spread influence, its epidemic nature, and its short duration. While it lasts it produces considerable prostration of strength and depression of spirits.

Etiology.—The facts about the various epidemics of influenza appear to be as follows. The affection arises with *extreme suddenness,* attacks *whole communities* within a few hours, passes rapidly from *country to country* and from *town to town,* is apparently *independent of the weather* and of the *seasons of the year,* attacks *persons of both sexes* and of *all ages.* The epidemics *rarely last longer than six weeks,* and during that time the majority of persons within their influence are attacked. The cause of the affection was formerly believed to lie in various abnormal meteorological conditions, some change in the electrical condition of the atmosphere, the presence of excess of ozone, and the like. It is, however, probable that influenza is due to a specific poison communicated in the breath from the diseased to the healthy person. The virus may be capable of indefinite multiplication, both within and outside the body. At any rate it is highly *communicable,* but is *not inoculable.* It is doubtful if an attack removes the susceptibility. There have been no distinct epidemics of late years. Sporadic cases are not uncommon, but are usually not severe.

Symptoms.—With or without a preliminary feeling of malaise, the patient is suddenly seized with a feeling of—

Chilliness or even with shivering, followed by all the symptoms of—

Catarrhal fever, including headache, a raised temperature, rapid and full pulse, tightness across the forehead, flushes, chills, sweats, pains in the limbs, watering of the eyes, running from the nose of a watery acrid fluid, stiffness, hot and disagreeable sensations about the throat, hoarseness and cough, tightness across chest and difficulty in breathing, furred tongue, loss of appetite, constipation, and scanty high-coloured urine of high specific gravity, depositing urates on cooling.

Prostration of strength and depression of spirits are some of the most marked features of the attack, and may increase to an alarming extent, muscular tremors and subsultus tendinum being occasionally observed, with apathy or delirium.

The attack lasts from three days to a week or ten days, when it abates, certainly leaving extreme weakness for days or weeks, and often leaving bronchitis and other grave sequelæ. Any inflammatory affection of the air passages, lungs, or pleuræ may complicate the course of the disease, *e. g.* pneumonia, pleurisy, bronchitis.

The catarrh of each of the districts of mucous membrane attacked passes through its usual course, but there is often much implication of the gastro-enteric tract, with persistent vomiting; jaundice and epistaxis are likely to occur in the course of the affection.

Prognosis.—Although influenza is not of itself a fatal disease, yet its influence in increasing the death-rate during an epidemic is remarkable. Its effects among a community are so wide-spread that the mortality from the affection among the old and debilitated is great absolutely, although the percentage mortality is small, about 2 per cent. only.

Treatment.—The patient must be kept in bed in a warm but well-ventilated room, and must be placed upon slop diet, milk, beef-tea, and good nourishing meat broth. The bowels should be opened with a mild purge (Calomel, gr. ij or iij), and a saline medicine, such as the Haust. Ammoniæ Acetatis, with Sp. Ætheris Nitrosi ʒss, and Tinc. Hyoscyam. ʒj should be administered; to this mixture some Oxymel Scillæ ʒss or Tr. Camph. Co. ʒj may be added if the bronchial catarrh is marked. Other expectorants and diuretics may be given. It is seldom advisable to bleed either generally or locally, but stimulant counter-irritation to the chest in the form of mustard and linseed poultices or Lin. Terebin. Aceticum is often a relief. In the epidemics of influenza it has, in nearly every case appeared best to administer alcoholic stimulants throughout the attack and during the period of convalescence.

Some object to **opium** in influenza, others depend upon it as upon a sheet anchor; it is, however, generally safe to use it, unless there is great pulmonary congestion and cyanosis. **Inhalations** of steam or of various **medicaments**, benzoin, conium, &c., are often of use.

After the severity of the attack has passed over, tonics, such as quinine, iron, and cod-liver oil should be administered.

The complications must be treated as they arise, after the plan followed when they occur as independent affections.

VIII. Hay Asthma, Hay Fever, Rose Cold

This affection is a peculiar acute catarrh of the respiratory mucous membrane, to which a minority of the community is subject in the spring and early summer, apparently induced by emanations from new-mown hay, or other odorous vegetable substances.

Etiology.—In the causation of hay asthma or hay fever we have two distinct elements, viz. predisposing and exciting; according to some authorities the former is a special irritability of the central nervous system, or of the respiratory mucous membranes, or of both. This, however, appears to be too general a statement, and we should be more disposed to accept Sir Morell Mackenzie's* plea that the predisposing cause of the affection is an idiosyncrasy, accepting at the same time his definition of that term, namely, that it is an exceptional state of the constitution

* 'Hay Fever,' 5th ed., p. 22.

shown in a few persons by a peculiar sensibility, not possessed by the bulk of mankind, to certain agents which manifests itself by certain definite phenomena. Towards the development of this idiosyncrasy, according to the same authority, the following circumstances conduce:

Race.—The Anglo-Saxon race appears to be alone liable, English and Americans being those who are almost exclusively attacked.

Temperament.—It generally occurs in nervous energetic people, of—

Fair social position, chiefly of the male sex; it is distinctly hereditary; occurs in early middle age, about forty, more than in the old or young.

The exciting cause in those predisposed to the complaint is now generally believed to be the pollen of the flowering grasses with which the air teems during the early summer. This pollen causes the irritation of the mucous membrane, which produces the phenomena of the disease. The pollen is most prevalent in the air at the end of June. It is asserted that it not only produces the local irritation, but also that it enters the circulation, and so sets up the constitutional disturbance. The pollen of the grasses is believed to produce most of the attacks, as it forms no less than 95 per cent. of the pollen found in the air; in meadows the sorrels, however, contribute a considerable amount. The pollen of the rose in some people produces a very similar affection to hay fever, hence the term Rose cold. In America another plant appears to be also implicated in the causation, viz. the roman wormwood.*

---

* Mackenzie (loc. cit.) figures the following grasses as those of which the pollen is most prone to produce the catarrhal attack :—(1) Sweet-scented vernal grass (*Anthoxanthum odoratum*), (2) oak-like grass (*Holcus avenaceus*), (3) fertile meadow grass (*Poa fertilis*), (4) meadow fox-tail (*Alopecurus pratensis*), (5) Rough-stalked meadow grass (*Poa trivialis*), (6) wood meadow grass (*Poa nemoralis*), (7) perennial rye (*Lolium perenne*).

Symptoms.—Sometimes the symptoms are those of an ordinary acute catarrh or coryza, when the conjunctivæ and nasal mucous membrane are affected, with more or less pharyngeal catarrh added.

Itching, smarting, and watering of the eyes with photophobia. Injection and sometimes chemosis of conjunctivæ and swelling of the eyelids.

Itching and running from the nose of a watery fluid, with swelling of the mucous membrane.

Sneezing; heat and dryness of the throat.

Pains in the limbs; some pyrexia, possibly urticaria.

The attack lasts for hours, or even days, according to its severity.

At other times, either with or without the above symptoms, there are the symptoms of acute catarrh of the bronchi, with spasm, as of an asthmatic attack. This lasts a longer or a shorter time, according to the duration of exposure to the irritation and its potency. The attacks closely resemble those of ordinary asthma but come on in the daytime, are as a rule less severe, and seldom last so long.

Treatment.—Those who are subject to this very peculiar affection must be as far as possible removed from the source of the irritation, and must avoid the country during the flowering season and remain in a town, taking care, however, not to fix upon a dwelling near a mews nor other place where hay is stored. A sea voyage during this period might be tried.

If the patient cannot remove from the country, a respirator, or thick veil of cambric, is recommended to be worn out of doors, or watch-glass "goggles" for the eyes.

Bathing the eyes with tepid water or with weak astringents relieves the irritation of the attack. Cocaine 4 to 6 per cent. solution is also advised.

Local application of cocaine spray to the nasal cavities, and the daily passage of a nasal bougie, are advised by Mackenzie, who believes medicated bougies do good (*e.g.* of gelato-glycerine 40 grms., hydrochlorate of cocaine $\frac{1}{10}$ gr., Atropiæ Sulph. $\frac{1}{220}$ gr., or of bismuth and acetate of lead).

Injections of quinine, **Ferrier's snuff,** smoking of tobacco or stramonium, Himrod's powder, &c., with arsenic, hydrocyanic acid, valerianate of zinc, belladonna, and other internal remedies, have all been recommended as beneficial in one or other form of the attack.

CHAPTER VII

DISEASES OF THE LUNG-TISSUE PROPER

In this chapter will be considered : i. Emphysema of the lungs; ii. Pneumonia; iii. Gangrene of the lungs; and iv. New growths of the lungs and pleuræ. The next chapter will be devoted to the subject of Phthisis Pulmonis.

I. Emphysema of the Lungs

Emphysema is a condition of the lung in which there is over-stretching of the pulmonary air vesicles with breaking down of the interalveolar septa, whereby a larger or a smaller number of vesicles, in different parts of the lung, tend to coalesce. In the common form, viz. *vesicular* emphysema, the mischief stops there, but in the other and rarer variety, *interlobular* emphysema, the air escapes from the vesicles and invades the interstitial and sub-pleural connective tissue.

Conditions under which it arises.—It is said that there is a considerable hereditary tendency to the affection and that this exists in quite half the total number of cases. The affection is found in men much more often than in women, chiefly because of more laborious occupations and other conditions of life which are peculiar to men. **Adults** are more affected than the young, and the tendency towards emphysema, as well as the severity of the disease, increases as age advances. As will be seen in the context, occupation may be a strongly predisposing cause, *e. g* playing upon wind instruments, occupations which necessitate straining to lift heavy weights, or shouting. As to the exciting causes, vesicular emphysema appears to be set up in acute bronchial catarrh, in chronic bronchial catarrh, in chronic pleurisy with adhesions, in phthisis and other disease of the lung-tissue, and in whooping-cough. Interlobular emphysema seldom occurs except in childhood. It arises from very violent paroxysms of cough in the course of any of the most frequent lung affections met with in childhood, viz. capillary bronchitis and whooping-cough.

The exact method of production of vesicular emphysema under different conditions is still a matter of disagreement. No doubt, however, different forces act in different cases. In the following epitome, an attempt is made to separate the cases of emphysema due to abnormal inspiratory pressure, from those due to abnormal expiratory pressure, and to establish a third group in which a change in the nutrition is the starting-point of the affection, and a fourth arising from rigid dilatation of the thoracic walls. It must, however, be at once confessed that this grouping is to a large extent arbitrary. In a considerable number of cases, it is almost certain that the cause is not simple, is not simply an increase of the inspiratory or expiratory pressure, but that where one is increased the other is diminished, and also that increase of pressure is associated with malnutrition and partial loss of elasticity of the lung-tissue, which becomes emphysematous. Emphysema may be either vicarious, when it occurs in certain portions of the lungs, to make up for other portions having become blocked up and impermeable to air, or substantive, and when the change is more general, and is not due to the pulmonary tissue being absolutely impermeable in any part.

Methods of Production.—1. Inspiratory emphysema.—Emphysema may be set up by immoderate and too protracted inflation of some of the air-cells, when from some cause there is obstruction to the free entrance of air into other parts of the lung. This coupled no doubt with more or less want of nutritive activity of the vesicles which become dilated, may occur:

(a) *In chronic phthisis*, when a considerable amount of the proper lung structure has become wasted or consolidated in large or small patches, provided the chest does not correspondingly shrink. The remaining air-vesicles dilate in consequence of the extra inspiratory pressure expended upon them, and the interalveolar septa may waste and break down.

(b) *In chronic pneumonia*, in which part of the lung-tissue undergoes consolidation, most often at the base, vicarious emphysema occurs elsewhere; similarly, we may suppose emphysema may be present in cancer and other conditions producing imperviousness of portions of the lung-tissue. It is extremely doubtful whether emphysema occurs in acute consolidation.

(c) *In chronic pleurisy*. When the lung is partially bound down by adhesions, so that the upper and posterior portions cannot expand freely, the anterior and lower borders may become emphysematous.

2. **Expiratory emphysema.**—Emphysema may be set up by excessive distension of the air-vesicles in forcible expiration, provided the cause act for a sufficiently long time. This condition is induced:

(a) *In bronchial catarrh*, in which the calibre of the air-tubes is diminished either by swelling of the mucous membrane or from accumulation of the secretion on the surface. There is more obstruction to the exit than to the entrance of air into the vesicles. This gradually causes accumulation of air in them, distension and finally emphysema.

(b) *In violent expiration with the glottis closed or*

contracted. In paroxysmal coughs, especially whooping-cough and catarrhal bronchitis, with spasm, as in asthma and from the occupation of playing upon wind instruments, and in other occupations before enumerated in which there is much muscular straining. The air is compressed in portions of the lung, having been at any rate partially expelled from others by the efforts of forcible expiration, the free exit of air from the thorax being at the same time obstructed or prevented at the larynx.

3. **Emphysema due to primary induration of the lung-tissue (atrophous emphysema)** is said to occur sometimes in old people in whom there is no primary mechanical defect in the entrance and exit of air; fatty degeneration of the pulmonary air-vesicles has in some cases actually been demonstrated. This primary malnutrition must not be mistaken for that which doubtless occurs secondarily to various lung diseases, more or less in every case of emphysema, and which is, as we have said, partially operative in its production.

4. **Emphysema secondary to permanent rigid dilatation of the thorax from ossification of the rib cartilages.**—Primary rigid dilatation of the chest has been asserted as a rare cause of emphysema. It seems unlikely that the mechanical cause could act alone. It is, however, quite reasonable to couple this with malnutrition of the lung as helping to account for emphysema in old people, without other apparent cause.

Interlobular emphysema is no doubt, as a rule, caused by violent expiratory efforts.

Post-mortem Appearances.—When the chest is opened the lungs do not collapse as is usually the case, but have a tendency to bulge, their volume being increased, except in the case of atrophous emphysema. The lung covers the pericardium on the left, and extends downwards on the right to the seventh rib when the condition is general.

In conditions of partial emphysema in the neighbourhood of the consolidated portions of the lungs, the distended vesicles vary in size from the normal up to that of a pea. Some remains of the interalveolar septa are mostly seen. These emphysematous bullæ are generally at the front and lower edge of the lung, but may be at apex or elsewhere.

In expiratory emphysema the upper part of the lung is that chiefly involved. In the other varieties the change may be found at any part. The lung-tissue is soft, either red in colour, or partly pale and mottled with black pigment; it is bloodless and dry.

In interlobular emphysema beneath the pleura here and there are seen air-bubbles, which can be moved about with the fingers, in lines bounding the lobule but not beyond. When these bubbles rupture during life, air is extravasated into the pleural cavity, forming pneumothorax; in other rare cases the air passes along the roots of the lungs into the areolar tissue of the mediastinum, and thence beneath the skin, producing subcutaneous emphysema.

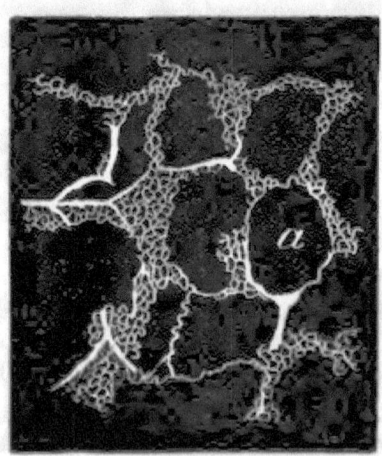

FIG. 32.—Pulmonary emphysema. An injected preparation. *a* is placed in a single infundibular dilatation arising from atrophy of the alveolar septæ. To the left of the figure and below are larger dilatations not single. (Ziegler.)

Symptoms.—Well-marked general emphysema may be present in a lung without producing any symptom except slight shortness of breath or puffiness, and as such is not uncommon in those otherwise healthy. This shortness of breath may be little felt as long as the patient is quiet and does not exert himself; on exertion, however, the breathing becomes more rapid and shallower than it should be. The alteration is chiefly marked with respect to expiration. The inspirations are of necessity rapid, because in an emphysematous lung the gaseous interchange is below the normal. The expirations are incomplete in spite of the fact that they become aided by the muscles of extraordinary expiration, and in consequence of this the air is incompletely removed from the lungs, or, in other words, the residual air is increased, so it comes to pass that the inspirations on occasions are more rapid and are at the same time more energetic, although the chest walls are capable of little movement. The muscles of extraordinary inspiration are pressed into the service in order that the thoracic capacity may be stretched to the greatest extent possible. There are several factors which enter into the production of the dyspnœa, viz. the loss of elasticity of the lung-tissue with the obliteration of the alveolar walls and their blood-vessels, and also the condition of permanent inspiration, as regards the chest wall, which has been already noticed. A considerable amount of bronchial spasm often intensifies the dyspnœa.

If the emphysema is exaggerated, and there is at the same time distinct malnutrition of the lung-substance, dyspnœa of considerable severity arises. When this is the case, the bronchial catarrh which probably preceded and was the primary cause of the affection, is much increased. As a consequence of this after a time the pulmonary circulation is seriously obstructed, with the result of producing:

Congestion of the mucous membrane of the bronchi, especially in the lower parts of the lung, with a tendency to œdema of the lung and hydrothorax in later stages.

Dilatation and hypertrophy of the right chambers of the heart, the dilatation gradually exceeding the hypertrophy, as the muscular fibres of their walls become the subject of degeneration. The pulse is small and weak, and may become irregular and intermittent.

Obstruction to the portal circulation, with its symptoms, congestion of the liver, gastro-enteric catarrh, dyspeptic troubles, possibly hæmatemesis and diarrhœa, enlarged spleen and ascites, enlarged abdomen, and even hæmorrhoids.

Renal congestion, with more or less albuminuria; scanty, high-coloured urine, of high specific gravity, depositing abundant urates on standing, and possibly blood and casts.

Cerebral congestion, with dizziness and headache, somnolence, or restlessness.

General cyanosis and dropsy, blueness of the face, with dilated capillaries about the cheeks and nose, blueness of the lips and tongue as well as of the hands and feet, with more or less œdema of the extremities and elsewhere.

If this condition progresses or persists, it must terminate in death, as increasing obstruction to the pulmonary circulation gradually produces so great a dilatation of the right chambers of the heart that the ventricular contractions, becoming more and more inefficient and incomplete, finally cease, congestions of the liver and other organs having, for some time previously, become more and more marked and troublesome. Incompetence of the tricuspid valve, with pulsation of the veins of the neck and of the liver, is sometimes noticed. Death is preceded by drowsiness, apathy, and coma.

Between the two extremes, emphysema with slight

dyspnœa, and emphysema which terminates fatally from the complications which arise during its course, there is every gradation of severity. It must not, therefore, be supposed that an emphysematous patient, even when affected with severe bronchial catarrh, is subject, of necessity, to the complications which have been enumerated.

The sequence of events in the history of an emphysematous patient is probably this: (1) Emphysema is produced by bronchial catarrh occurring in a lung not quite healthy. (2) The emphysema tends to increase the catarrh. (3) Obstruction to the pulmonary circulation, by the destruction of the alveolar walls in places, and with obliteration of their capillaries, as well as by the delay in the circulation from the bronchial congestion, tends, after a time, to produce dilatation of the right heart. (4) Dilatation is followed, or accompanied, by hypertrophy. As long as the hypertrophy is compensative, the emphysematous patient is subject only in a slight degree, if at all, to the inconveniences caused by the congestion of the internal organs, even although he suffers from a certain amount of bronchial catarrh in the winter. As the years pass on, and the emphysema increases with them, however, the winter bronchitis comes on earlier and is more severe, and the tendency is for the congestions of internal organs to become more and more serious, indicating the advance of the cardiac dilatation, until at last, in spite of treatment, the heart, the muscular fibres of which have been gradually atrophying and degenerating, is no longer able to carry on the too greatly obstructed circulation, and the patient dies.

It is practically impossible to say how much of all this is due to emphysema alone, and how much to the accompanying bronchitis.

Physical Signs.—The general appearance of a patient suffering from chronic bronchitis and emphysema is almost characteristic. His face is swollen and bluish, complexion is muddy, conjunctivæ watery, vessels of face dilated, alæ nasi projecting, lips and tongue bluish, neck is thick and short while the sterno-mastoids, and other muscles, which raise chest, are prominent and cord-like; the jugulars are dilated.

On examination of the chest by—

Inspection.—It is found to be large and barrel-shaped, with the ribs and sternum raised. There is scarcely any appreciable movement on inspiration. The attachment of the diaphragm is mapped out on the chest by a line of dilated capillaries. On coughing, the supraclavicular regions swell out, producing the so-called emphysematous tumour. This is sometimes very marked. It is due to the dilatation of the apex of the lung and also to distension of the jugular vein and sinus.

Palpation.—The vocal fremitus is diminished. The apex-beat of the heart is scarcely perceptible, but there is, very likely, epigastric pulsation.

Percussion.—The resonance over the whole of the chest is greater than normal, more tympanitic than subtympanitic, and the areas of cardiac and hepatic dulness are diminished or absent.

Auscultation.—The respiration is either very feeble, or the inspiration is somewhat exaggerated. Dry crackling crepitation or crepitus may sometimes be distinguished. In the normal district of bronchial breathing, viz. the interscapular region, this latter sound may be absent. If bronchial catarrh be present, rhonchus, sibilus, and moist crepitations may be audible.

The heart-sounds are heard very indistinctly over the cardiac region proper, except the pulmonary second sound, which may be distinct and even accentuated. In the epigastrium the sounds are sometimes audible enough.

The symptoms and physical signs of **vicarious** emphysema, if it exists to a sufficient degree to produce any, are much the same as those of the substantive emphysema we have been discussing.

Interlobular emphysema, unless it travels to the subcutaneous tissue of the neck and elsewhere, producing the so-called surgical emphysema, which crackles on pressure with the finger, gives rise to no characteristic symptom or physical sign.

Prognosis.—Emphysema of itself is not dangerous. It is so closely associated with bronchitis, however, that it is almost impossible to consider them apart as regards prognosis. The affections may ultimately produce death, but they may go on conjointly with periods of quiet for many years. Death may also result from complications at any time, *e.g.* from acute bronchitis or pneumonia, sudden cardiac failure, &c. It is also said to have occurred during an asthmatic attack.

Treatment.—Emphysema in a slight degree requires no treatment, nor indeed, as a rule, does vicarious emphysema, but precautions should be taken to prevent increase, and so bronchial catarrh should be carefully guarded against, and should be energetically treated if it arises.

Thus an emphysematous patient should avoid catching cold, should wear flannel next his skin both in summer and winter, should stay at home in wet or cold weather, and should, if possible, seek a warm climate during the winter and spring.

Certain specific methods of treatment have been recommended for emphysema, some of which it is necessary to mention, viz. (1) the employment of compressed air or of oxygen for the dyspnœa, whereby the respirations are diminished in number. The elasticity of the lungs is said thereby to be aided, the congestion of the mucous membranes diminished, and the return of the venous blood to the heart increased in consequence of the increased pressure upon the vessels within the thorax. The experience of this method in this country does not appear to support the high praise bestowed upon it as a curative measure by Continental physicians. (2) Compression of the throat has been tried with some success, but the risk of hæmorrhage and convulsions appears to be considerable. (3) A residence among the pine woods is strongly believed in by good authorities, and the treatment is reasonable.

As regards **complications**,—the various remedies recommended for bronchial catarrh would be indicated if that affection became pressing. Iodide of potassium is particularly good for the asthmatic attacks; œdema of the lungs, and a greatly overcharged pulmonary circulation, may call for bloodletting; dropsy may require purgatives and diuretics, and hydrothorax may require paracentesis. As a rule alcoholic stimulants should be given, especially if the patient is old or debilitated.

II. Pneumonia or Inflammation of the Lung Substance

Anatomically there are three varieties of pneumonia:

- A. **Croupous Pneumonia,** an acute affection, commencing in the lung-tissue proper, attacking entire lobes, hence called **lobar,** and characterised by a fibrinous exudation, which fills and distends the air vesicles and air tubes of the affected part.
- B. **Catarrhal Pneumonia,** a subacute or chronic affection, associated with and mostly secondary to bronchial catarrh, attacking smaller districts of the lung, hence called **lobular,** and in which there is accumulation within the air vesicles chiefly of young epithelial cells, with little or no fibrinous exudation.
- C. **Interstitial Pneumonia,** a chronic affection always secondary to some lung or other disease,

attacking the alveolar walls as well as the interalveolar and interlobular connective tissue of the lung.

A. **Croupous or Lobar Pneumonia.**—Inflammatory consolidation attacking whole lobes of the lungs and characterised by an exudation into the pulmonary alveoli and air tubes of an exceedingly fibrinous material occurs under a variety of different conditions. Sometimes it is the result of injury to the lung-tissue, when it is called **traumatic**; sometimes it occurs in the course of the acute specific fevers and of other affections, when it is called **symptomatic** or secondary, but most commonly it constitutes a part of a special disease, which is called **acute lobar pneumonia** or **pneumonic fever**. Before considering this special affection, to which our attention will be chiefly devoted in the present section, it will be as well once for all to mention in greater detail the circumstances under which traumatic and secondary pneumonic consolidation respectively arise.

(a) **Traumatic** consolidation may occur (i) from direct injury to the lung, e.g. from fractured ribs; penetrating wounds of the thorax, including gunshot wounds; blows; foreign bodies in the air passages; and (ii) from inhalation of hot air or of other irritants.

(b) **Secondary** consolidation may be (i) *symptomatic* of other affections, occurring in (1) Fevers, e.g. scarlet fever, variola, enteric fever, diphtheria, pyæmia, and the like; (2) Bright's disease of the kidney, especially in the amyloid variety of the affection; (3) Acute rheumatism. Or it may arise (ii) *by extension, e.g.* in (1) Croup of the larynx; (2) Pleurisy; (3) Pericarditis.

Acute lobar pneumonia, a special disease, is apt to occur *in any part of the world,* and scarcely appears to be more prevalent in one climate than another, although it is generally considered to be more *especially a disease of temperate climates*. *Season of the year* exercises a decided influence over its prevalence and its fatality, but this differs in different countries, *e. g.* the largest number of cases of pneumonia in this country occurs in the winter from December to February: on the Continent, however, in spring and early summer, but in both early autumn, August and September, furnish the fewest cases. In this country November and December are the most fatal months, whereas on the Continent March and May are. The fewest deaths from the disease occur in September and August. Speaking generally, two thirds of the cases occur in winter and spring, and one third in summer and autumn. Neither severe cold nor the amount of atmospheric humidity appears of necessity to increase the mortality from the affection. A *sudden change in the weather* from heat to cold (and the reverse?) increases the number of cases at once; so also do *easterly or northerly winds* with moderately cold weather.

The influence of epidemics and other diseases upon the number of pneumonic cases appears to be very slight, and a marshy malarial soil is without constant influence of any kind. *Dwellers in towns* are more likely to be attacked than country communities, and those *who live an indoor sedentary life* are more liable than those whose occupations lead them out of doors.

The affection may attack the individual *at any age*, the young or the old. Men are more frequently attacked than women, and in them the disease is more fatal. In an analysis of 985 cases of acute lobar pneumonia treated in St Bartholomew's Hospital from 1876 to 1885 inclusive, 689 were males and 296 females. The largest number of cases occurred between the ages of 15—30. Of the total 182 patients died, viz. 134 males and 48 females.

Although pneumonia may run a more typical course in the strong and robust, yet it is more liable to attack those who are *weak and debilitated* from any cause, but especially from *alcoholic excess*. Dusty occupations do not seem to produce pneumonia, nor yet playing upon wind instruments nor singing.

As for the **exciting causes,** it is generally admitted that a *chill* is not infrequently the starting-point of pneumonia, but how much influence it has in producing the attack is uncertain. It is as impossible to deny that some cases of acute lobar pneumonia do apparently arise from a chill of some sort, as it is to deny that others occur without any such exciting cause.

The proportion of the cases in which there is a distinct history of a chill is about 25 per cent. It has been implied that pneumonia is *endemic*, and that it occurs in places regularly about the same times of the year. There is accumulating evidence, based upon good authority, to show that it may be also *epidemic*, affecting whole communities, that it may attack one after another *the members of a household*, and that it may be *communicated* directly or indirectly *from the sick to the well*. Occasionally the epidemics are apparently *connected with bad drainage, sewer gas*, and *overcrowding*, and with the affections such as enteric and typhus fever and dysentery, which are usually associated with such conditions. One attack of pneumonia does not remove, but rather increases the susceptibility to the disease.

As to **the nature of the disease**, two different views are held, the *first*, that acute lobar pneumonia is essentially a local inflammation of the lung due to exposure to cold, the attendant fever being secondary to the local inflammation; the *second*, that the affection is a constitutional one, with the lung inflammation as the local manifestation of the general disease. The evidence in favour of the constitutional nature of many cases of the disease is very strong. It may be thus summarised:

1. Acute pneumonia bears a close resemblance to the acute specific fevers, both in its sudden onset and in its termination by crisis.

2. The symptoms of fever usually precede the local signs of pulmonary inflammation by many hours or even days.

3. The severity of the fever is not in proportion to the extent of the local inflammation, being sometimes great when the local lesion is small and *vice versâ*.

4. That the affection occurs as an epidemic, that it attacks members of the same household one after another, and that it is in all probability capable of being communicated directly from the sick to the well.

5. Micro-organisms of a specific character have been discovered in the sputum, exudations of blood of many patients suffering from the affection which have been said to produce, by inoculation into suitable animals, a similar disease.

It will be generally agreed that this evidence is sufficiently strong to allow us to consider that acute pneumonia is a constitutional affection in many instances. Can we go a step further and say that it is always so? Hardly so at present we think, as we are at once confronted with the fact that in a large number of cases of highly typical pneumonia, beginning suddenly and terminating by crisis, exposure to cold immediately preceded the onset, being the only apparent cause.

If the disease is a constitutional one, always or only sometimes, the virus which produces it, is in all probability, a specific micro-organism. But how this micro-organism acts, whether locally or primarily by means of the blood or lymph infection, or sometimes by the one means and sometimes by the other, is not certain.

Micro-organisms.—In the foregoing account it

will be noticed that we have not spoken with such certainty about the micro-organisms as some would do. This is not due to any want of belief in the power of micro-organisms to produce the disease, but simply because at present the results of experiments have been somewhat contradictory. Many observers have demonstrated, however, in the affected lungs, in the sputum, and in the blood of patients suffering from acute lobar pneumonia, special *micrococci* or *diplococci*, which were described by Friedländer as of oval shape, occurring singly, in dumb-bells, in chains, or in zoogloea. They seem to disappear after the crisis of the fever. Cultivations of these micro-organisms, pneumococci as they have been called, in meat broth, meat extract solution, and in nutritive gelatine, when injected into the subcutaneous tissue of rabbits or white rats, are stated to produce pneumonia. It is not admitted by all, however, that the disease produced by the inoculations of the cultivations of pneumococci into rabbits and mice is true lobar pneumonia, but rather that it is a septicæmia " due to a specific septicæmic micrococci not neces-

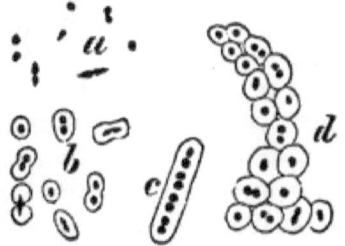

Fig. 33.—Diplococci of pneumonia (Weichselbaum, Fränkel).—*a*, Cocci without envelope; *b*, single- and double-celled cocci with gelatinous envelope; *c*, chain of cocci with envelope; *d*, colony of cocci.

sarily always present in the sputum and lungs of primary croupous pneumonia" (Klein, 'Micro-organism and Disease,' Ed. ii, p. 74).

Thus we see that the exact relation of the pneumococci to the disease is not yet clearly defined.

Pathology and Morbid Anatomy.—Croupous pneumonia is ushered in by hyperæmia followed by inflammation of the mucous membrane of the air passages, with exudation into the latter of a highly fibrinous material. It generally attacks the lowest lobe, and more frequently the right lung. The apex, however, is every now and then the part first attacked; this is especially the case when the subject is debilitated from alcoholic excess or old age. The inflammation and its precedent hyperæmia are believed to spread downwards from the larger tubes to the air-vesicles. The blocking of the air passages, however, occurs in the reverse order. It has a tendency to spread from lobe to lobe, so that the whole lung may become consolidated.

Three stages in the process of the lung consolidation are recognised. They are as follows:

I. **The first stage, engorgement or splenisation.** Although two processes mark this stage, viz. hyperæmia and commencing effusion, they cannot well be differentiated. The lung in this condition is heavier than natural, and the affected part is of a deep red or reddish-brown colour. The lung is œdematous, pitting on pressure with the finger and retaining the impression like an œdematous limb. On section the cut surface looks red, and when squeezed exudes a frothy, somewhat viscid, sanguineous fluid, and breaks down more easy than natural. In this last particular as well as in its colour, it somewhat resembles the spleen tissue, hence the expression splenisation. Portions of the affected tissue will float if thrown into water, and will also crepitate when squeezed between the finger and thumb.

Microscopically, the pulmonary arterioles are full of blood, as are also the capillaries.

Here and there are small extravasations. In the pulmonary alveoli a variable amount of material, consisting of blood-corpuscles, both red and white, epithelial cells, and a certain amount of fibrin fibrils—as yet, however, the fibrin fibrils are in small amount. No doubt during life these materials are suspended in serous fluid.

II. **The second stage** of exudation, called, from the resemblance the lung bears in this condition to the liver, **red hepatisation**. By this stage the fibrinous exudation which had commenced in the former stage reaches its maximum, filling up the lung-tissue and rendering it quite solid.

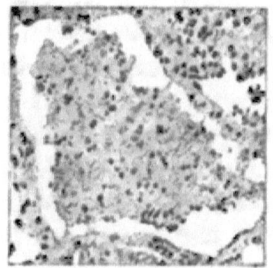

FIG. 34.—A section showing an alveolus of the lung filled with fibrinous exudation, in the second stage of croupous pneumonia. From a coloured drawing (× 100).

The solid lung is heavy, weighing, it may be, twice or three times as much as in a healthy condition; inelastic, more or less swollen, marked by the impression of the ribs, and deep red in colour. On section the cut surface is red, marbled or mottled with lighter spots, and granular, does not crepitate, and only exudes a scanty amount of red serous fluid when squeezed. The solid tissue sinks in water, and although firm easily breaks down.

Microscopically, the air vesicles are seen

to be closely packed with masses of fibrin fibrils, enclosing within their interlacing meshwork blood-corpuscles, chiefly colourless, and epithelial cells and débris. Here and there evidences of rupture of small vessels are to be made out from hæmorrhage into the vesicles and their walls. In addition to the exudation into the alveoli themselves, there is generally an increase in the connective-tissue corpuscles of their walls, and a variable number of leucocytes into and about the sheaths of the smaller arteries.

III. **The third stage** of purulent infiltration is called also the stage of **grey hepatisation** from its characteristic appearance. The lung is still heavy and sinks in water, is yellowish grey in colour, and on section the coarsely granular red look is exchanged for one more finely granular, pale grey, or greyish yellow. The tissue is exceedingly friable, pits on pressure from breaking down, and the insertion of the finger produces an irregular cavity, which on the removal of the finger is at once filled with yellowish pus.

The solidified lung may or may not crepitate on squeezing.

Microscopically, the vesicles are seen to be less packed than in the second stage. There is less fibrin, but the corpuscles are much increased in number (pus-corpuscles), and although the coloured blood-corpuscles have disappeared, cells are undergoing fatty degeneration. In the inter-vesicular tissue the appearance of leucocytes (and proliferating connective-tissue corpuscles?) is more marked, but the pulmonary capillaries are less gorged with blood.

It should be recollected that in addition to the conditions, which were enumerated on page 211, in which lobar pneumonic consolidation proper occurs, there are others in which a consolidation may be found very like it. They are the cases of heart, lung, and kidney disease in which the bases and dependent parts of the lung are affected with what is known as hypostatic congestion due to interference with the pulmonary circulation in those parts. This condition may or may not pass into a low form of inflammation, but whether there be inflammation, or not, the pulmonary tissue is found solid from an exudation of plasma and blood-corpuscles into the alveoli. It may be, therefore, that this consolidation should be treated of under the head of catarrhal pneumonia.

The pleura, over the affected part, in cases of acute lobar pneumonia, is nearly always the seat of acute inflammation, and is covered with thick exudation. A variable amount of fluid effusion into the pleural sac may also occur, but not so frequently. If the effusion is great, the physical signs of the pneumonia may be more or less masked.

Resolution may take place during any one of the three stages, but most usually occurs in the second. The fibrin and cells undergo fatty degeneration, the contents of the alveoli liquefy, and are either expectorated or absorbed; and a similar fate may befall the pus-corpuscles when resolution takes place in the third stage. They are either expectorated or, having undergone caseation, are absorbed. In both cases, however, as distinct interstitial inflammatory changes have commenced in the interalveolar tissue, some time must elapse before complete recovery can take place. This is, one would imagine, especially true in the cases where there has been over-stretching or rupture of the air vesicles.

The interstitial inflammation, however, may continue or even increase, and so be the beginning of the more chronic affection of the lung, which is sometimes a sequel to croupous pneumonia, viz. cirrhosis or fibroid induration.

Sequelæ of croupous pneumonia.—Of the more uncommon pathological complications of croupous pneumonia may be enumerated the following:

α. Abscess. This occurs when the inflammatory process has been very acute, or more intense than usual. It is very rare. In it there is direct breaking down of the lung-tissue itself, from over-pressure of the infiltration upon the alveoli and blood-vessels, and the small abscesses formed in this way may coalesce and form larger ones. These abscesses may (1) ulcerate into neighbouring bronchi, (2) ulcerate into pleural sac, (3) caseate or calcify, and in the former case, give rise to phthisis.

β. Gangrene. A rare termination, which occurs from the sudden cutting off of the blood-supply from considerable tracts of tissue by the pressure of the inflammatory exudation.

γ. Phthisis from part of the exudation undergoing caseation, and not being expectorated or absorbed; or from the fibroid induration mentioned above.

Symptoms.—In a typical case of croupous pneumonia the onset of the disease is sudden. In the cases in which there is a distinct history of a chill, symptoms begin in from twelve to thirty-six hours, after exposure to the exciting cause. In the majority of cases there is a distinct—

- **Rigor**, generally a single and prolonged one, which lasts for half an hour or more; sometimes shivering of less distinct character, more rarely a feeling of chilliness only. One or other of these varieties is seldom absent, in fact, in less than 10 per cent. of the cases.
- **Convulsions** sometimes take the place of the rigor, especially in children, and occasionally, it is said, although we should think very rarely, there is complete loss of consciousness.

Vomiting is also sometimes the initial symptom.

Very soon after one or other of these symptoms the patient presents all the symptoms of—

High fever. He complains of *severe headache*, often frontal, dizziness, and pains in all of his limbs. The *temperature is high* and quickly runs up to 104° or 105° F., and in severe cases a degree or two higher. There are daily remissions; the temperature is lowest in early morning, and highest in the afternoon.

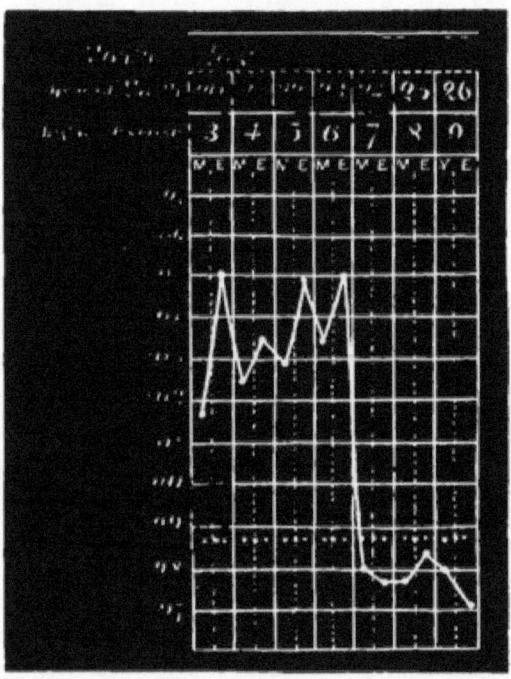

Fig. 35.—Temperature chart of a case of acute pneumonia, which terminated by crisis on the seventh day.

The skin is dry and pungently hot. As the fever advances the *disturbances of the digestive tract are very marked;* loss of appetite, great thirst, thickly coated dry tongue (sometimes very brown and cracked); sometimes nausea and vomiting and constipation, very rarely diarrhœa.

The *pulse is small and feeble, quick and rapid,* the beats varying in number, in adults, from 80 to 120.

The face is flushed, and the flush is generally most marked on the same side as the inflammation of the lung—a deep reddish-blue flush; the eyes are shining and bright, and the lips are often bluish in serious cases.

To these symptoms of fever there are added:

Difficulty in breathing, which is usually very marked. It comes on quite early in the case and remains throughout its course, possibly increasing day by day. The patient has to be *propped up in bed*, and his appearance is almost characteristic of the affection. His *anxious and pained expression*, with his *rapid and shallow respirations*, amounting to 30, 40, or even 50 per minute, his *alæ nasi dilating* at each breath, all point to the nature of his malady. The dyspnœa or orthopnœa is due not only to fever, it would appear, but also to the great diminution of the respiratory surface. The pulse and respiration ratio is deranged, and is altered from 4 to 1 to 2 or 3 to 1.

Pain in the affected side, which is due in the majority of cases to the attendant pleurisy. This pain is often felt at the outset of the attack, and is, as a rule, only temporary.

Cough, too, begins early in the course of this affection, generally within twelve hours from the rigor. It is at first short and dry, and then with slight frothy mucous expectoration. When the affection is well established, unless the patient is too young or too feeble to spit, there is—

Sputum, of a highly characteristic nature. It is very viscid and rusty, so tenacious as to adhere to the containing vessel even when it is turned over. In the earliest stages it may contain a considerable amount of blood, when its colour is bright red, sometimes yellow streaked with blood. In unfavorable cases later on in the affection it is deep reddish purple or like prune juice. Under the microscope are seen many blood-corpuscles and cells of all shapes, some of them undergoing fatty change, and sometimes fibrinous casts of the small tubes; chemically it is found to be rich in chlorides, but contains no alkaline phosphates.

Herpes about the lips, and less frequently about the nose, cheeks, and eyelids. This eruption is not absolutely characteristic of pneumonia, as it sometimes occurs in other acute fevers, but it is considered to be an important diagnostic sign when it does appear, which is not by any means invariably the case. By some physicians it is believed to be a sign of good omen.

Changes in the urine.—The urine is diminished in quantity, of increased specific gravity, high coloured, and the solids, especially the urea and uric acid, are increased in quantity. The chlorides are absent or greatly diminished, at any rate after the first stage of the disease. Albumen is frequently present, and sometimes blood and fibrinous casts. When the crisis occurs the chlorides reappear in normal or excessive amount. The total urea eliminated during the fever is rather below the normal amount.

In addition to the foregoing symptoms, which are almost certain to make their appearance in a well-marked case, there are several others sufficiently frequent to be classed as symptoms of the disease; they are:

Delirium. The headache and sleeplessness may pass into delirium, especially in childhood or old age, in the anæmic and debilitated, and in the excitable. It is especially prone to arise when pneumonia attacks the apex of the lung instead of the base, which is its usual locality. In persons addicted to alcoholic excess, pneumonia is frequently attended with the symptoms of delirium tremens from its commencement, but the delirium may not occur until much later than this in other cases.

Jaundice. This symptom is present in a not

inconsiderable number of cases. It is not slight as a rule. It is often attended with moderate enlargement of the liver. Its cause is often obstruction to the bile-duct from catarrh, or congestion of the portal circulation, but it may occur without any such palpable cause.

Course.—From the first attack, the symptoms of fever continue for some days without intermission. The temperature oscillates between 103° and 105° F., sometimes reaching or even passing 106° F., but not doing so as a rule. The dyspnœa increases as the area of the inflammation of the lung extends. Anxiety of countenance continues. Sleep and rest are interrupted by attacks of coughing which are very troublesome, no doubt partly from the adhesive nature of the expectoration. The thirst, anorexia, and red dry tongue continue. The pain in the side generally abates after the first day or two. The patient is evidently very ill. On one of the days, between the fifth and the ninth, generally on the seventh or eighth, the crisis occurs in the fever. The temperature in a few hours drops from 105° F. or higher to 97° F.; the pulse also becomes less frequent, the dyspnœa is relieved, cough is less troublesome, and the sputum is much less glairy, in fact all the symptoms are relieved. The patient may go off into a critical sleep and wake up sweating profusely, or he may have a critical discharge of a large amount of urine. In the cases in which such a marked crisis occurs, the local inflammation is not long in clearing up and convalescence is established.

In some cases *the crisis is delayed* until the end of the second week, the symptoms in the meantime taking on an almost typhoid character; but then somewhat suddenly improvement sets in. Again, there may be a *gradual, instead of sudden, fall of temperature*, or a sudden fall to a certain point, followed by a further gradual fall to normal.

Symptoms of asthenia are marked in the old and

feeble and also when the disease is complicated with gastro-enteric catarrh. In these cases many if not all of the characteristic features of the affection may be absent. Recovery generally occurs before the stage of purulent infiltration has been reached.

Physical Signs:
- On inspection, the movement of the affected side is distinctly impaired, especially at the base. The apex-beat of the heart is not displaced.
- On palpation, the vocal fremitus over the affected part is distinctly increased.
- On percussion, there is at first slight, if any, impairment of resonance, running into complete dulness later on, with somewhat increased resistance. The dulness does not alter on change of the position of the patient. The upper limit is a curved line following the direction of the fissures of the lungs. The dulness does not extend beyond the border of the sternum on the affected side.
- On auscultation, at first the almost characteristic pneumonic "crackle" is noticed on inspiration, followed by tubular breathing becoming bronchial, with bronchophony (sometimes pectoriloquy or ægophony), and when resolution is taking place, crepitatio redux, a larger and moister râle than the original crackle. Above the affected part of the lung and in the unaffected lung the breathing is puerile.

Terminations.—I. Recovery. In ordinary cases there is a decided tendency towards recovery, which frequently takes place as above mentioned by crisis. As regards the epidemic form, the severity varies. On an average the death-rate from this affection may be put down as 1 in 5 or 6.

II. Incomplete recovery, leaving as the four chief sequelæ those which have been before mentioned in the description of the morbid anatomy, viz. α. *Abscess.* β. *Gangrene of the lung.* γ. *Cheesy infiltration.* δ. *Fibroid phthisis.*

III. **Death,** which may arise directly from **asphyxia** from interference with the oxidation process in the lung, especially when the whole of one lung is affected or parts of both lungs; or from more or less sudden hyperæmia and collateral œdema. It may also occur, from **asthenia,** in those debilitated from any cause —old age or alcoholism; from complications; or from **hyperpyrexia.**

Complications.—The complications which may arise in the course of **acute pneumonia** are so numerous that, as has been well said by an able writer upon the subject (Jurgensen), "a complete statement of them all would have to embrace the whole of pathology." The following, however, are the most common: *pleurisy with effusion, bronchitis, pericarditis, endocarditis and meningitis, rheumatic fever, and inflammation of the parotid.* This list is exclusive of *abscess* and *gangrene of the lung, caseous pneumonia,* and *fibroid phthisis,* which have been already mentioned.

Prognosis.—Among unfavorable indications may be mentioned the following:—i. Old age or extreme youth. ii. The male sex. iii. Debilitated conditions, especially from alcoholism. iv. Very high temperature, above 106°, or a pulse of over 120, respirations over 50. v. The existence of other diseases or supervention of complications. vi. The absence of sputum. vii. Delirium. viii. Prune-juice sputum. ix. Typhoid symptoms. x. Severity of the character of the epidemic.

The favorable indications would be the reverse of the above conditions, but in addition it may be mentioned that proof of the localisation of the pneumonia to the lower lobe, with no tendency to spreading, would be, as a rule, a sign of good omen.

Treatment.—In addition to the treatment of symptoms, which is generally similar in all cases, we must proceed in the treatment of acute lobar pneumonia according to the condition of the patient, the severity of the inflammation, and the height of the fever. Thus from the beginning of the case the broad lines to be followed can, as a rule, be indicated, and it is generally possible to determine early whether the case is likely to require active treatment, either of a stimulant or depressant nature.

(1) **In ordinary cases**, that is, when the affection occurs in healthy adults, unless the general or local symptoms be unusually severe, exceedingly little treatment is required, as the tendency is towards recovery. Under such circumstances it is only necessary to place the patient under favorable dietetic and hygienic conditions, and to attend to the due performance of the bodily functions. Thus the patient should be kept in bed in a well-ventilated, not too warm (temp. 60° F.) room, and should be fed upon slop diet, such as milk, arrowroot, beef-tea, and the like. A purgative at the beginning of the attack, such as calomel (gr. iij—v) or castor-oil (ʒj) is certainly a good thing, and the bowels should afterwards be carefully attended to and assisted by gentle purgatives, if necessary, to regular action. A *saline draught*, such as Liq. Ammoniæ Acetatis and water, or a solution of chlorate of potash (gr. xv), may be given three or four times a day. A large "jacket" *poultice* of linseed meal should entirely encase the affected side, and should be changed frequently. We have had no experience of the utility of *cold compresses* applied to the side, but many consider them as useful as, if not more so than, the poultices.

(2) **Serious cases**, whether serious from the debilitated condition of the patient, from old age,

from alcoholic excess, from the presence of other diseases or complications, from the severity of the constitutional fever, or from the extent or position of the local inflammation, require *stimulants* from the first. It will be seen that these constitute a large proportion of the total number of cases. Stimulants, therefore —*dietetic, alcoholic, and medicinal*—should be administered. Thus strong beef-tea, meat extracts, soups, chicken broth, and milk should be given frequently, say every half or three quarters of an hour, in small quantities at a time, and brandy or wine should be freely administered. Medicinal **tonics** and *antipyretics*, such as quinine and the like, and *stimulants*, *e.g.* carbonate of ammonia, are also indicated.

(3) **Plethoric cases**, or those in which the patient is young, robust and lusty, may possibly require a modification of what is called the *antiphlogistic* or depressing treatment. This was at one time the only treatment of pneumonia. It consisted in the abstraction of blood by venæsection frequently repeated, until enormous quantities had been removed, and the exhibition of tartar emetic to an adult, together with low diet and purgatives. Such a treatment, unless considerably modified and mitigated, *cannot be recommended in any case*. It is still a question whether general bleeding is ever indicated, but if it be done at all, it should be done as Dr Gairdner recommends: (*a*) In very early stages; (*b*) in uninjured constitutions; (*c*) when the fever is high or the symptoms are urgent, provided that there is no indication of exhaustion. "Blood-letting (in pneumonia) should be reserved for severe cases," he adds, "diagnosed from the urgency of the fever, pain, dyspnœa, and the general strength of the patient."

Symptoms which may require treatment:
1. **Pain or stitch in the side.**—Never use blisters. Apply a few leeches, dry cups, or a hot poultice with a little mustard in it.
2. **Want of appetite.**—In the treatment of this very important symptom the mineral acids, nux vomica, quassia, or calumba, do good.
3. **Thirst.**—Ice to suck (which is often contra-indicated by the dyspnœa), lime-juice or lemonade, or saline drinks, barley water, toast and water.
4. **Cough.**—Should not be stopped, expectorants are not of much use; if extremely troublesome and preventing sleep a small quantity of a tincture containing opium may be tried. Try first of all, sedative inhalations of conium and the like.
5. **Insomnia.**—Try to do without opium, but if this be found impossible administer it in the form of injection of morphia ($\frac{1}{12}$—$\frac{1}{4}$ gr). This may also relieve pain and cough.
6. **Delirium.**—Usually requires alcohol and good feeding, ice to the head, and in some cases opium or Tr. Cannabis Indic. ♏xx cum Mucilagiæ ʒj; Aquam ad ʒj o.n.; or pill of the Ext. gr. ij cum Ext. Hyos. gr. ij o.n.
7. **Dyspnœa.**—Dyspnœa, when urgent, is sometimes improved by venæsection.
8. **Hyperpyrexia.**—Digitalis, quinine, aconite and the like. Sponging the body with tepid water, or cold baths, may sometimes be required, but if the temperature be very high, above 105° F., as a rule the lung mischief is very extensive, and so the treatment of hyperpyrexia by cold bathing is not so hopeful as in some other affections in which it may occur without such extensive local disorganization.

After-treatment.—After the fever abates, unless some complication arise, the progress of the patient towards recovery is generally without interruption. Good diet, tonics, *e. g.* quinine or the mineral acids and cinchona should be exhibited, and in many cases cod-liver oil (ʒij bis die) or iron is of distinct benefit. Convalescence should be followed, if possible, by change of air.

B. Catarrhal Pneumonia (Broncho-pneumonia, lobular pneumonia)

Causes.—This affection, which is very seldom anything but a subacute or chronic disease, is always associated with, or preceded by, catarrh of the bronchial mucous membrane. It results from catarrh of the bronchi, either directly by extension of the inflammation downwards, or by inhalation of the inflammatory bronchial secretion, which sets up inflammation in the vesicles; or indirectly, by first of all producing collapse of a portion of the lung, which is followed by congestion and inflammation. Any cause, therefore, which gives rise to bronchitis, including measles, whooping-cough, typhoid fever, and the like, may produce catarrhal pneumonia; and again, any cause which interferes with the natural depth of the inspiration may tend towards collapse or so towards catarrhal pneumonia.

Morbid Anatomy and Pathology.—The appearances which the lung presents in catarrhal pneumonia depend to some extent upon the circumstances which have produced it.

a. When the inflammation has been secondary to collapse of the lung following upon bronchitis, the affected parts are to be found chiefly in the lower and posterior edges of the lung, although it is possible that entire lobes may be in the same condition. The collapse appears in the course of bronchitis, which is due to partial blocking of the bronchial tubes, and to the weakening of the inspiratory power. The collapsed parts are not of necessity confined to the positions above indicated, but when small may be scattered irregularly through both lungs. The collapsed portions of the tissue are dark bluish, depressed below the general surface, and easily break down. The consolidated patches vary in size from a line in diameter to a quarter of an inch. They are never very hard, are of a reddish grey passing to grey or yellow, not distinctly marked off from the surrounding tissue, and on section appear more or less granular. Later on, the consolidated patches become harder. The bronchus leading to the affected part

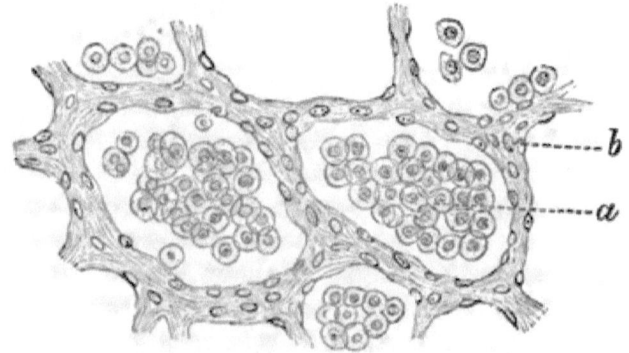

FIG. 36.—Catarrhal or Broncho-pneumonia. Two complete, and other incomplete, alveoli filled with epithelial cells, *a*; the walls, *b*, of the alveoli are somewhat infiltrated.

is usually congested and contains much glairy mucus. On microscopic examination, the alveoli are found to be more or less filled with small cells or inflammatory corpuscles and epithelial débris, but there is no evidence of fibrinous exudation. It is strongly insisted upon by some pathologists that in the majority of cases the appearance of these consolidated patches is no proof of inflammation in the collapsed districts. They believe that the contents of the alveoli are partly derived from the congestion set up by the collapse, and partly by inhalation of the bronchial secretion. This may indeed explain the origin of the majority of the consolidations, but the presence of the inhaled bronchial secretion within the alveoli would in all probability set up inflammation secondarily.

b. Catarrhal pneumonia proper, *i. e.* that which is dependent upon collapse of the lung, either occurs in very young children, generally under three years of age, or in old people. In the former case, as a result of measles or whooping-cough, and in the latter, following in the course of chronic bronchitis. It may occur, but is rare, in adults. It can easily be imagined that in measles and like affections the acute catarrhal inflammation attacking the bronchi, even to their minutest ramification, may spread to the pulmonary alveoli, and that an inflammation of the alveoli may arise from the inhalation of the inflammatory bronchial secretion. The parts affected by this inflammation are recognised post mortem, by presenting small consolidations varying in size from that of a pea to that of a hazel nut, scattered irregularly throughout both lungs, slightly raised, of colour varying from brownish red to greyish yellow, somewhat granular on section. As a rule quite evidently lobular, but sometimes so closely set together in a lobe as to give it appearance of lobar consolidation, which is only to be understood by careful examination. A little fluid exudes on squeezing. Microscopically, the same appearances are found as in the first variety. My own experience has shown me that in many cases of consolidation of the above kind there is little evidence of inflammation, although the alveoli may be packed with epithelial débris. The condition, however, is not seldom conjoined with distinct inflammation of the alveolar walls.

Patches near the surface of the lung, consolidated by catarrhal pneumonia, are frequently covered by pleuritic exudation (Fagge).

As regards the terminations of the anatomical condition; the alveolar contents appear to undergo fatty degeneration, and the number of exudation- or pus corpuscles increases, and so the offending material is got rid of by expectoration. The other terminations are much the same as in croupous pneumonia, viz. (α) **caseous infiltration** and (β) **abscess** (rare). There is no evidence of **gangrene** arising from this condition, or of **cretation** or **calcification**, but the latter, at any rate, is very possible.

Symptoms.—The symptoms are very indefinite, as the affection is never primary. The chief indications upon which reliance is to be placed to indicate that the capillary bronchitis has penetrated into the alveoli are:

- **Alteration in the character of the cough.** The child evidently dreads to cough from the pain it gives him, and in whooping-cough the prolonged fits of coughing are succeeded by a short hacking.
- **Increase in the symptoms of fever**, shown by elevation of temperature, say from 102° F. to 105° F. or more, with hardly any variation for several days; rapid pulse, up to 150 or over, and extremely rapid respiration.
- **Extreme restlessness**, tossing about, turning from side to side, or requiring constant shifting. Increase of dyspnœa and failure of heart's action.

Physical Signs.—The physical signs also indefinite; sometimes *patches of dulness* may be made out, first of all along the spine behind, on either side, when collapse of the corresponding portion of the lungs has preceded the inflammation, but as a rule the patches of consolidation are not sufficiently large to alter the percussion note, especially if they are set in the midst of resonant lung. If the collapse and consolidation increase beyond a certain point the physical signs cannot be distinguished from those of croupous pneumonia. It is occasionally possible to map out *small areas of impaired resonance with harsh (bronchial) breathing and bronchophony. Crepitation* rather less fine than in croupous pneumonia, is also said to be audible, with inspiration and expiration.

Course and Terminations:

1. When the **disease** takes a favorable course and ends in **recovery** the fever does not suddenly terminate in crisis, but gradually abates by lysis, and even when the fever is gone, convalescence is usually gradual. Catarrhal pneumonia seldom fails to leave the lung-tissue in a more or less damaged condition.

2. When the **recovery is incomplete** the sequelæ are, as above stated, much the same as in croupous pneumonia, but caseous infiltration is more common. Emphysema and even dilatation of the bronchi also occur.

3. **Death** is comparatively frequent but is seldom sudden. (a) In some cases, however, death may occur in a few days in feeble children, with all the symptoms of acute asphyxia. (β) Generally death is preceded by marasmus, and the pneumonic patches are found to be in a state of caseation. This form is most frequent when the affection is secondary to whooping-cough or chronic bronchitis. It is believed to constitute one form of pulmonary consumption.

Diagnosis.—In children the difficulty of diagnosing catarrhal from **croupous pneumonia**, unless the case has been watched from its commencement, may be very great. Some criteria are to be obtained, firstly, from the absence of dulness or impaired resonance, or if either be present, from its being irregularly distributed on both sides; and secondly, from the gradual as opposed to the sudden abatement of fever.

From tubercular meningitis.—Occasionally children affected with catarrhal pneumonia over and above the whooping-cough or measles, develop symptoms almost exactly similar to those of tubercular meningitis. In fact, as both tubercular meningitis and catarrhal pneumonia with head symptoms, occur as sequelæ of measles, it is almost, if not quite, impossible to distinguish between them. In both conditions there may be apathy, boring of the head into the pillow, muscular twitchings, vomiting, squint, and convulsions.

Treatment.—Nothing is specially indicated in the treatment of catarrhal pneumonia. The administration of emetics is believed in by some, supposing there be an **evident accumulation** in the bronchial tubes. Poultices to enclose the chest or stimulating liniments, *e. g.* Lin. Terebinthinæ appear to be comforting; cold compresses are said to be equally efficacious. In any case it is advisable to use, with care, alcoholic stimulants. The administration of a few drops of brandy with each ounce of milk is often of great benefit. Considering the close connection which apparently exists between the occasional after-effects of this affection and pulmonary consumption, the convalescence of the patient should be watched with the closest attention.

c. **Interstitial Pneumonia** (Fibroid induration, fibroid phthisis, cirrhosis of the lung) consists in the gradual increase of the fibrous tissue of the lung from chronic inflammation. It is always **chronic** and always **secondary** to some other affection of the lungs, bronchi, or pleura. It is generally **localised**, but sometimes affects whole lobes.

    i. It is associated with certain trades in which dust and irritating particles are constantly inhaled, and is therefore especially found among miners, coalheavers, knife-grinders, cotton-spinners and the like.

    ii. It is a sequel of croupous or catarrhal pneumonia, as already indicated, especially of the latter.

    iii. It is associated with pleurisy, as it is apt to occur in a lung which has long been rendered airless by pressure of pleuritic effusion.

iv. It arises in the neighbourhood of any consolidation of the lungs, e. g. tubercle, cancer, and the like. It often acts in localising these deposits.

v. It is associated with chronic bronchitis, and bronchiectatic cavities.

Morbid Anatomy and Pathology.—On post-mortem examination of a lung in which interstitial pneumonia has been well established, the affected portion is solidified and airless, and is generally pigmented; it cuts almost like cartilage, and is dry and firm; white bands are seen running through the section. Little remains of the proper lung-substance. The bronchi are frequently found dilated or pouched.

In coniosus pulmonum or miner's lung the tissue of the lung and of the nearest bronchial glands is cirrhosed and marbled black, but it is said that the pigmentation need not of necessity be associated with pneumonia or fibroid induration of any kind. Such simple pigmentation, however, one would imagine to be uncommon.

The process consists of an inflammatory hyperplasia of the connective tissue of the lungs, and the changes take place, (a) in the alveolar walls; (b) in the interalveolar tissue; and (c) around the bronchi (peribronchitis). The new fibroid growth is at first vascular, but when it contracts it causes encroachment upon the alveoli and the blood-vessels are obliterated; in some cases (d) the contents of the alveoli undergo fibrous development.

Under the microscope may be seen, in the situations indicated, a fibrous tissue with a varying number of cells, sometimes of fusiform shape.

The pleura over the affected part of the lung is invariably more or less thickened.

Symptoms.—As interstitial pneumonia is essentially chronic and secondary to other affections, presenting few if any symptoms which can be distinguished from those of the primary disorder, most information is to be derived from physical examination of the chest. When bronchiectasis has been definitely established, however, some symptoms arise (p. 186) which are of importance.

In those whose occupation has been the cause of their trouble, the symptoms begin with gradually increasing shortness of breath, cough, very gradual loss of flesh and anæmia, with occasional night-sweats.

The chief symptoms to be feared in the course of this disorder are those which point to dilatation of the right heart, uncompensated by hypertrophy. When this happens, cyanosis and dropsy with œdema of the lung may after a variable time be succeeded by death.

Physical Signs:
    On inspection, there is flattening and general contraction of the chest, with considerable diminution in movement.
    On palpation, the vocal fremitus may or may not be increased; the heart apex-beat is displaced, frequently pulled towards the affected side.
    On percussion, there is dulness and increased resistance.
    On auscultation, there is feeble breathing, together with more or less crepitation on cough, and if bronchiectasis exists, cavernous breathing and pectoriloquy.

The diagnosis between a **bronchiectasis** and a phthisical **cavity,** it is often extremely difficult, if not impossible, as the physical signs are practically identical; reliance must be placed upon a careful examination of the symptoms and course of the affection, and of the locality of the physical signs in the chest.

    Treatment.—No special form of treatment is indicated for this affection. It is essentially chronic. The symptoms which require attention are chiefly the cough and the sputum. The former should be treated by **stimulant expectorants**, the latter may be diminished, and also made less fœtid by **carbolic acid or creasote inhalations**. We always employ **counter-irritation** in these cases either by blistering fluid, or by means of iodine applied over the affected parts. **Tonics**, *e.g.* strychnine, iron, and cod-liver oil, are decidedly useful. To the **operative treatment** of bronchiectasis attention has been already directed.

III. Gangrene of the Lung

Gangrene attacking the lung-substance is a somewhat uncommon condition, which is found under the following circumstances:

1. In **acute pneumonia**, it is an occasional complication from sudden arrest of the circulation; it also occurs in the violent inflammation which arises from the entrance of foreign bodies into the air passages.
2. In **chronic diseases of the lung**, *e.g.* phthisis, interstitial pneumonia, with or without bronchiectasis, from more gradual obstruction of the small vessels from thrombosis, associated with feeble cardiac action.
3. From **embolism**, especially when the embolus is of infective material; thus it occurs sometimes instead of pyæmic abscess.
4. In **debilitated people and drunkards** sometimes.

It may be *circumscribed* or *diffused*; the former is more common; the spots vary in size, they are deep bluish green, moist and extremely fœtid; they seem to have a tendency to shell out, and are abruptly defined; they are situated at the surface and chiefly in the lower lobes; when diffuse a whole lobe may be attacked.

Symptoms.—If in the midst of one of the above-

mentioned conditions the patient begins to have an intensely—
> Foul odour of the breath and foul-smelling sputum, which may be blackish grey, and separates into layers depositing a sediment similar to that from bronchiectasis,

it is very probable that gangrene has been superadded to the primary affection.

Physical Signs.—The physical signs may not be different to those of the affection in the midst of which gangrene has arisen. There may be the signs of a cavity with or without dulness on percussion.

Prognosis.—Very unfavorable, although not absolutely fatal to life; some patients succumb very shortly after the condition sets in; a few seem to be affected by it but little, at any rate at first.

Treatment.—Purely palliative and symptomatic. In order to mitigate the offensiveness of the breath, inhalations of carbolic acid or creasote may be employed and sumbul or musk may be administered internally.

IV. Cancer of the Lung and Pleura

Cancer is, comparatively speaking, a rare disease in the lung; when it occurs it is more often secondary to cancer elsewhere than primary. The form of cancer is most commonly medullary, but it may be scirrhus or alveolar, and sometimes it is with difficulty distinguished from sarcoma. The tumours may be single or multiple, they vary in size from a mustard seed to that of a small orange, and may occupy the whole of one or more lobes of the lung. Some difference in their mode of growth depends upon the way in which they originate, for example:

1. They may spread inwards from the mediastinum from the bronchial glands, and then the growth is found to be in connection with the bronchi.

2. They may originate in the lung-tissue; when they mostly form one mass.
3. They may originate in and about the pleura; being either pedunculated or sessile.
4. They may be secondary to disease elsewhere, and then are mostly multiple; especially to cancer of the breast in women, and of the testicle in men.

The disease is nearly always confined to one lung, and more often the right than the left. The tumours are white as a rule and are often umbilicated and exhibit a softened centre. Sometimes the tissue breaks down to form cavities, and sometimes portions undergo caseation.

Symptoms.—*a. When primary*, the symptoms are very obscure, but probably include some of the following:

Pain, well marked and persistent, sometimes localised very definitely.

Anæmia with the peculiar complexion of malignant infection.

Emaciation, possibly not marked until the disease is far advanced.

Dyspnœa, sometimes apparently such as occurs in extensive bronchitis.

Short cough with frothy mucous expectoration when the bronchi are implicated, mixed with blood or with the so-called "**red-currant jelly sputum.**"

Hectic fever.

The affection may be latent for a time, or may be mistaken for **bronchitis**, which is nearly always present if the new growth is in the neighbourhood of the bronchial tubes, but it is much more likely to be mistaken for **pleurisy** or **empyema**. The latter mistake is made all the more possible, as the physical signs closely resemble that condition, and very frequently the chest is tapped with the idea of drawing off the fluid. It should be remembered too that the new growth, if it be in the neighbourhood of the pleura, is nearly certain to set up pleurisy, and if this be the case the diagnosis becomes still more obscured.

Physical Signs.—The signs depend, of course, upon the position and extent of the disease, but as soon as the tumours are sufficiently enlarged to give definite physical signs, they are much more like those of pleurisy with effusion than phthisis or other lung consolidation, or than bronchitis. If the growth bulges the chest wall here or there, or if it erode the ribs, a light is thrown upon the diagnosis.

b. *When secondary*, the lung implication is very likely to be overlooked, unless there is such manifest evidence of the disease in the axillary, bronchial, or other glands, as shall suggest a frequent and careful physical examination of the thorax.

Prognosis.—Refers only to the duration of life. Death usually arises from complications, especially pleurisy. The duration of the cases is well under a year.

Treatment.—Simply palliative and symptomatic.

CHAPTER VIII

PHTHISIS PULMONALIS

Consumption, Pulmonary Consumption, or Wasting Disease of the Lung, Tubercular Disease of the Lung, &c.

The exact signification of the term **phthisis pulmonalis** or phthisis is still a matter of some difference of opinion. In its most extended meaning it is taken to include all the diseases of the lung in which there is loss of substance preceded by consolidation and softening or breaking down of its tissue, arising from any cause whatever, whilst in a more restricted sense it indicates a constitutional disease, the local manifestation of which is the eruption of tubercles in the lung. Of the many other interpretations of the term, it is only necessary to mention in this place that it is used sometimes, in the present day, to signify any disease of the lung due to the presence within its tissue of a specific micro-organism, the so-called *Bacillus tuberculosis* of Koch.

Whatever view, however, is taken of the pathology of phthisis, to the consideration of which we shall return presently, it will be generally admitted that cases of phthisis differ materially in their modes of origin, symptoms, course, and terminations, as well as in the post-mortem appearances of the lungs. It will be as well, therefore, to begin our description of the disease with an account of the conditions under which consumption of the lung arises, using the term in its most extensive and popular sense.

Modes of origin.—It is exceedingly difficult, if not impossible, to gauge the value of each of the many circumstances which appear to have to do with the causation of phthisis, in determining the disease. Some circumstances, indeed, which in one case may be merely predisposing, may in another be directly exciting and *vice versâ*. It will be readily understood, therefore, that in bringing together the various supposed causes of phthisis, no attempt will be made to place them in order of gravity, or to specify the exact part each of them plays in the production of the disease. One exception, however, will be made. Phthisis pulmonalis is a disease which so very frequently runs in families, that hereditary or family predisposition as a cause of the affection requires special and careful consideration, and will be treated of first of all.

Hereditary or family predisposition.—The fact that in a very large number of the cases of consumption one can obtain distinct evidence that the affection had previously attacked some member of the patient's family cannot well be ignored or minimized.

Such a family history can be obtained in one third, at the very least, of the total number of cases, and probably might be obtained in a still larger proportion. The family transmission of the disease is most often *direct*, that is to say from fathers to sons and from mothers to daughters, or it may be *crossed*, that is to say from fathers to daughters or from mothers to sons. It may also be *indirect*, from uncles or aunts to nephews or nieces. Sometimes the hereditary taint appears to *miss an entire generation* and is received from a grandfather or grandmother. In some cases whole families are carried off by this disease, or the sons or daughters, one after another, on reaching a certain age, develop its symptoms, and in each of them it may often pass through much the same course and terminate after much the same time. If both parents are phthisical, the hereditary tendency naturally is much increased. One meets with singular instances of phthisis carrying off all the members of a large family, except one or perhaps two individuals, who may seem to be particularly strong and healthy.

Such in brief are the facts with respect to hereditary transmission of the disease. The explanation of the working of this cause in producing the disease is even now only partially understood. There are several theories upon the subject. (1) That the germ of the disease itself is handed from father to son or from mother to daughter seems to be very unlikely, but this idea was formerly held. (2) The most widely accepted view of the matter appears to be this, that phthisical patients may beget strumous children. Struma is a condition in which there is a tendency to low forms of inflammation. When strumous inflammation attacks internal parts it is characterised by a tendency to produce exudations difficult of absorption, and which are consequently prone to undergo fatty degeneration or caseation. So a pneumonia attacking the apex of the lung of a strumous person may lead on to phthisis, by producing caseous consolidation of the tissue, or when attacking a lymphatic gland may result in a chronic enlargement from caseation of the exudation which has taken place therein. But even this explanation can scarcely be called satisfactory. It only shows how an inherited delicacy of constitution acts; it does not suggest in what it really consists. (3) Some physicians consider that the hereditary taint is simply a weakness of constitution whereby the individual becomes ill from slight causes, and in whom one or other part of the body again and again is subject to attacks of disease. If the part repeatedly attacked be the air passages, as by catarrh, phthisis may follow in due course. (4) Even those who hold that all phthisis is due to the tubercle bacillus are unable to ignore the part played by hereditary tendency in the predisposition to the disease. Green, for example, says that speaking generally it is simply a tendency to disease which may be said to consist in some feebleness of the constitution in general, and often of the lungs and other organs in particular.

Supposing then an inherited tendency to disease of the lungs, with or without the additional tendency to inflammation of a low type which is so strongly insisted upon by Niemeyer and others of his school, the question naturally follows, *why are the apices attacked more frequently than other parts of the lungs?* Inherited weakness of constitution no doubt has some part in producing this special vulnerability of the apices to the disease, but the immediate cause of the phthisical lesion in the apex appears to be a local one. The respiratory activity of the lung is less in the apex than elsewhere even in health, but in certain conditions of the chest the deficiency becomes very marked.

These conditions are either congenital or acquired. They are (1) contraction of the upper part of the chest wall, either from early ossification of the cartilage of the first rib, and subsequent imperfect development of the expansion of the chest, as was long ago suggested by Freund, or from occupation, habits of stooping or what not; and (2) imperfect development of the inspiratory muscles or loss of muscular power from any cause producing the alar or phthinoid, and flat chests (p. 84). Thus imperfect expansion, from the shape of the chest, and from want of muscular power (both of which defects may be congenital), diminishes the inspiratory expansion of the apices of the lungs. To this must be added firstly that they are more subject to atmospheric pressure than other parts, since they project out of the thoracic cavity above the clavicles, and secondly that they are subject to expiratory pressure during forced expiration. All of these anatomical points—and there are others which might be mentioned, mark out the apices from other parts of the lungs—in phthinoid chests. *What are the effects of their diminished inspiratory movement upon the apices?*

1. The secretions of the parts, or foreign matters taken in by inspiration, *e g.* dust and micro-organisms, are not so easily got rid of, and are therefore more prone there than elsewhere to set up irritation and inflammation. Similar substances expelled from the lower parts of the lungs may be by forcible expiration driven upwards into the apices.

2. The circulation through the part may be diminished and sluggish, and congestion, stasis, and possibly hæmorrhage may result.

We may now sum up the matter thus:

Inherited tendency appears to act not only by producing a delicacy or weakness of constitution, whereby the individual is unable to resist even slight attacks of illness, but also by producing a tendency to inflammation of the skin, mucous membrane, and lymphatic tissues of a low type, resulting in caseous consolidations in various tissues, especially in the lungs and bronchial glands. In addition to this, by interfering with the due development of the chest and its muscles, it specially leaves the apices of the lungs liable to be attacked by such chronic inflammations.

Although the family predisposition to consumption is great, there is no doubt, on the one hand, that the disease may arise in those who have no family predisposition, whilst, on the other hand, it is obvious that some with a marked inherited tendency escape the affection. In the former case it would seem that a delicacy of constitution has been acquired or that the chest has been in some way or other so altered locally as to be liable to the attack of phthisis, and in the other one would expect that some exceptionally favorable conditions have been sufficient to counterbalance or to keep latent the inherited tendency.

Before considering the question of acquired delicacy of constitution, its varieties and causes, it will be as well to mention several points with regard to the distribution of phthisis.

Age and Sex.—Women affected with consumption are more likely to be able to give some family history of the same complaint than men. In other words, phthisis is *more likely,* as the numbers affected of each sex are about equal, *to be acquired in men* than in women, without hereditary predisposition. According to Dr Reginald Thompson the sexual distinctions of acquired phthisis may be deduced from statistics to be the following:

Males.	Susceptibility.	Females.
Intensified after fifteen years of age, at its acme between twenty and thirty-five.		Intensified after ten years of age, loaded upon the period between twenty and twenty-five.
	Form of the disease.	
Subacute.		Acute in nearly one half of the cases.
	Fatal cases.	
Fewer fatal cases within eighteen months.		More fatal cases within eighteen months.
	Copious hæmoptysis.	
Frequent.		Less frequent.

The *male hereditary cases* closely follow the characters of the female *acquired cases,* that is to say, the susceptibility is especially marked between the ages of fifteen and twenty-five, and the acute cases as well as the fatal cases are more numerous. The tendency to copious hæmoptysis is not, however, increased; as regards the *female hereditary cases* the susceptibility of the early period of life is increased between ten and twenty-five. The fatal cases, the acute cases, and the cases of hæmoptysis are all increased as compared with acquired cases; the acute cases by one half.

Race.—Certain races of mankind are undoubtedly more liable to phthisis than others. No very definite information has yet been formulated upon the subject, but, at any rate, the *coloured races* appear to be more prone to the disease than white men. Contact with civilization almost invariably increases the tendency to consumption among savage or semi-savage peoples.

For example, the Indian settlers in the neighbourhood of the various trading stations of the Hudson's Bay Company are simply decimated by the disease, whereas the whites are little if at all affected. Of course there are many possible explanations of these facts. This special proneness may possibly be due to evil habits acquired from their neighbours, to alteration of the customs and the substitution of a less free life for one passed entirely in the open air. But even after all these elements of causation have received due consideration the remarkable fact yet remains.

Climate and locality.—Phthisis may and does exist everywhere in the world, but it is certain that in some climates cases are more frequent than in others. In *high altitudes phthisis is remarkably uncommon*, the number of cases diminishing as the height above the sea level is increased. To quote Dr Andrew,* we may say that "the influence of climate, whatever it may be, so far as it depends upon latitude, is slight and constantly modified or overridden by other more potent conditions. Further, it is probable that phthisis is relatively as well as absolutely more common in the tropics than in the extreme north, but this cannot be looked upon as having been yet certainly established."

As regards locality, by the researches of Buchanan in 1866—1867, certain very distinct relations were demonstrated between the prevalence of phthisis in the counties of Surrey, Kent, and Sussex, and the

* 'Lumleian Lectures,' 1884.

condition of the soil may be thus summarised. That there is less phthisis among those living on pervious soils than among those living on impervious; less among those living on high-lying pervious soils than low-lying pervious soils; less among those living on sloping impervious soils than among populations living on flat impervious soils. That draining the soil considerably reduces the phthisis mortality. The conclusion arrived at by Dr Buchanan, which may now be affirmed generally, and not only of the particular districts first reported upon, is, that wetness of soil is a cause of phthisis to the population living within its influence. This conclusion is all the more plausible from the previous researches of Dr Bowditch in America on the same subject, since his deductions were similar.

As regards age, sex, race, climate, locality, and the like, it is difficult to determine the exact manner of their action in predisposing to the disease. There are certain circumstances, however, which appear to act either in producing a delicacy of constitution, such as we have already spoken about, or which act locally in diminishing the proper respiratory movements, especially of the upper part of the lungs. Many of them act in both of these ways. Some of these predisposing causes may now be mentioned.

Bad hygienic conditions.—These may be thus summarised: want of fresh air, want of healthy exercise, want of cleanliness, and want of proper food. They may act together or separately, but are generally conjoined. On the other hand, even when they are all present they may not produce phthisis.

Occupation.—There are certain occupations strongly predisposing to phthisis, in some of which a considerable number of accompanying unfavorable conditions, both general and local, are present. Thus (a) working indoors in a confined space, and breathing impure air, is likely to be much less healthy than an active life in the open air. (*b*) Any occupation involving working in a cramped position, stooping, pressing against the chest, as in the case of compositors, cobblers, and the like, which locally interferes with natural and healthy respiration. (*c*) Any occupation which involves sudden alternations of temperature from heat to cold, or the reverse, are apt to engender catarrhs of the respiratory mucous membrane, and so indirectly to produce phthisis. (*d*) Any occupation which involves breathing a dusty or otherwise irritating atmosphere, *e. g.* knife- and scissor-grinders, flax-dressers, stonemasons, coal-miners, polishers, grocers and the like, not only predisposes to, but probably also excites phthisis.

Other diseases.—Many cases of acquired phthisis appear to date their commencement, from an attack of fever, such as typhoid, smallpox, measles, or scarlet fever, or from syphilis. The former affections act by producing a debilitated condition of the system, the latter generally in the same way, but sometimes by actually attacking the lung-tissue itself. Those who suffer from diabetes mellitus are predisposed to phthisis.

Over-lactation, rapid child-bearing, and excessive mental depression, especially when joined with any other unfavorable circumstances already enumerated, *e. g.* want of air and food, undoubtedly may have their share in producing excessive debility and proneness to the disease.

The whole of those conditions which have now been considered probably act chiefly, if not entirely, as predisposing causes only. Those which we shall now consider probably act as *exciting* causes.

Infection.—The idea is generally widely held by the laity in many countries that phthisis can be communicated by those affected with the disease to those who are unaffected, by direct contagion, the channel of communication being their exhalations. A similar view is also upheld by a very respectable minority of the medical profession. It no longer admits of doubt that one form of phthisis, viz. the tubercular, is capable of being propagated by inoculation. The researches of Villemin, Lebert, Simon, Burdon Sanderson, and others, have proved beyond a possibility of error that tubercle can be produced in animals by inoculations of phthisical sputum, of caseous masses from the lungs, and other phthisical material, while the more modern observations of Koch, Watson Cheyne, and others, have rendered it highly probable that the infective part of the inoculated material consists of a micro-organism of a special nature, now called the bacillus tuberculosis. Thus the propagation of one form of phthisis by inoculation is quite possible, and there is, as far as we see, nothing at all unreasonable about the idea that phthisis may be communicated from the diseased to the well, but we must confess that at present the clinical evidence in its support is unsatisfactory and insufficient; even if phthisis is ever propagated in this way, the cases occur very infrequently.

It is worth while to give the results of the investigation upon this point as conducted by the committee of the British Medical Association, published in 1884.

The Collective Investigations' Committee reported that they had issued the circular containing the questions relating to the possible communicability of phthisis on two occasions to every member of the British Medical Association with the following result, asking for the opinion of each individual member as to whether phthisis is contagious, and for evidence based upon his experience for or against this view. One thousand and seventy-eight (1078) answers were received. Of this number 673 were simple negatives, but the remaining answers contained observations of some kind which were classed as follows:

Class 1. Affirmative observers, 261. Class 2. Doubtful observers, 39. Class 3. Negative observers, 105. Total 405.

Of those 261 who answered in the affirmative the history of the cases they described showed 158 instances of apparent infection *between husband and wife;** 81 *between members of the same family, i. e.* between parents and children, brothers and sisters, &c., as well as between husband and wife; 13 referred to cases of supposed contagion between *persons unrelated, as well as between members of the same family;* and 8 were cases observed between *persons unrelated* only.

In going over those of the cases which are epitomised in the appendix to the report, it must be confessed that many of them rest upon evidence somewhat unsubstantial; at the same time it must be admitted not a few appear to be well supported. It is certainly a remarkable fact that so many as over 261 medical men in this country believe that they have seen cases of phthisis communicated from one person to another.

Looking upon the specific bacillus as the exciting cause of many cases of phthisis, and remembering that in the close and confined atmosphere of a single and often small room, a wife may be practically shut up with a phthisical husband whose every breath contains the specific organisms, one cannot but see that we have ready to hand the materials for, and the condition likely to produce, a communication of the disease from the husband to the wife, and in addition to this the wife is probably pulled down in health, by nursing and by her endeavour to provide food for the family, and, it may be, comforts and medical

---

* In the interesting and able report compiled by Dr Burney Yeo upon the result of this inquiry, it is pointed out that family predisposition and communication from one to another are not antagonistic if they appear together as causes, but may be related, the first as predisposing, and the latter as exciting.

attendance for her invalid **husband.** On the other hand, the extreme rarity of the reported cases (and those who believe in the **communicability** of phthisis at **all events must be anxious to report all** the **cases which** support their views), **as well as** the strong opinion in the negative by those who have had many years of special hospital experience, of chest complaints, **must suggest,** as before observed, that if it ever be a **cause of phthisis,** communication of the disease from **one person to** another is an extremely infrequent one. **As far as** my experience goes, of many years' work **at a consumption hospital,** it must be said that no **single case of** undoubted communication of phthisis from **the sick to the well** has come within my knowledge.

The bacillus tuberculosis is believed by some **to** be the **sole** exciting cause of phthisical lesions.

Exposure to wet and cold must be mentioned as being in **some cases** the apparent cause, at any rate, **of phthisis.** This is reasonable when we consider that such exposure **may set up catarrh** of the respiratory membranes **from which may follow caseous** infiltration of the lung-tissue.

Occupations in which **dust and** other irritating matters are constantly inhaled, evidently may **act** both as exciting as well as predisposing causes.

Pneumonia, especially of the chronic catarrhal **form, pleurisy, bronchitis,** and all those affections in which caseous material is left behind in the tissues **and organs of** the body, but especially in the lungs, must be considered as the immediate **cause of** the lung lesions of **phthisis.**

Hæmorrhage into the pulmonary alveoli.—It is probable that blood when retained within the pulmonary alveoli is an occasional **cause of** pulmonary consumption. It acts as an irritant foreign body, **and** sets up inflammation in its neighbourhood. The products of the inflammation and the retained blood afterwards undergo cheesy metamorphosis, soften, break down, and produce loss of lung-substance.

Excessive exertions of the body, which produce congestion of the lungs with increase of cardiac action, are stated by some to excite consumption. Niemeyer's note on the subject is interesting, although it must be confessed that the suggested connection may not be, as he affirmed, in the relation of cause and effect. "I possess among my reports of cases a number of examples in which excessive dancing or similar exercises were immediately followed by the first signs of a commencing pulmonary consumption, without any possibility of cold having taken place at the same time." He also suggests that to this class belong cases in which pulmonary consumption is said to have been caused by drinking cold water when the body was heated by exercise.

Morbid Anatomy and Pathology.—In a lung from a case of chronic phthisis various appearances are presented in different districts of its tissue, which illustrate not only the clinical stages of the affection, but also to a considerable extent show many of the different lesions of the disease. Let us therefore consider in order the various post-mortem appearances.

The pleura.—On opening the thorax, the affected lung does not collapse naturally, and is seen even from the front to be covered by a very **thick pleura** (Fig. 42). On endeavouring to remove the thoracic viscera the lungs are usually found more or less adherent to the chest wall, especially in the neighbourhood of the apices, and it often happens that the lung-substance is lacerated in the process of removal. Adhesions to the diaphragm, too, are not uncommon.

Solidification.—When removed from the thorax the greater part of the lung is found to be more or less solid, except possibly at the extreme base. On section, the consolidation is seen to occur in patches, varying in size from that of a mustard seed to that of a hazel nut or walnut, and also varying in colour from a white or grey hue to yellowish green; they vary also in consistency. These patches are irregular in shape as well as size, but many of them, especially the small ones, are of a more or less rounded form; some appear like small bunches of grapes (Fig. 43) attached to small bronchi as their stalks. Occasionally whole lobes are found quite solid. Many of the patches are softened in the centre, and some of them appear to consist of thick pus.

Cavities.—Cavities or vomicæ are almost certain to be found, especially at the apex and also laterally about the middle lobe on the right side. The cavities vary in size from that of a pea to that of an orange. Sometimes the whole of the upper lobe is occupied by a single large vomica, less often the whole lung is excavated, leaving little of its tissue, but more frequently a series of irregular cavities leading into one another extend from apex to base. The walls of these cavities are sometimes thick and their interior smooth and level, with a distinct membrane; at other times, the walls are imperfect and the interior rugged and rough, presenting a worm-eaten appearance. Then again the cavities may be full of pus, or may contain but little of that fluid. The larger ones are nearly certain to communicate with a bronchus.

Fibrous growth.—The part of the lung-tissue which is not occupied with the consolidation is generally firmer than natural, and often there are to be seen firm and thick bands of fibrous tissue running across the lung structure from one side to the other, sometimes acting as septa separating the cavities from other portions. In the cavities themselves are often to be seen bands passing from side to side, and not infrequently mistaken for obliterated branches of the pulmonary artery, but being most often fibrous bands only.

Pigment.—An abnormal amount of pigment is not infrequently noticed in the lung-tissue.

Enlarged bronchial glands.—The bronchial glands are nearly always enlarged, and frequently pigmented.

Tubercle of pleura.—On the surface of the lung, but evidently in the pleura, are often seen minute greyish-white dots, which are found on microscopical examination to be grey tubercles.

Cretaceous masses.—Here and there in the lung-tissue the knife in cutting grates against some hard substances, and these on examination prove to be cretaceous nodules.

In addition to the morbid appearances in the lungs the heart is usually small and atrophied; the liver fatty or fibroid, sometimes pigmented; the kidneys are not infrequently fibroid; the larynx and intestines ulcerated; the brain anæmic, sometimes exhibiting signs of tubercular meningitis.

Such then in brief are the post-mortem appearances to the naked eye of a case of chronic phthisis. We will next turn to the—

Microscopical examination of the various lesions. The consolidated masses, be they large or small, are made up in one of the following ways:

i. **Catarrhal pneumonic patches.**—On section of a patch of catarrhal pneumonia the outer zone is seen to consist of pulmonary air-sacs filled up with epithelial cells, and exhibiting the usual appearance of a lung so affected. At the outer margin of the zone the air-sacs are not so full as elsewhere, and so the disease passes gradually into the healthy tissues. The next more internal zone of a patch also exhibits the same character of contents, but here the air-cells are more distended and the epithelium cells are seen to have become more granular, while some have broken down into a granular detritus. More internally still these changes in the vesicular contents have become more marked, the epithelial cells having lost their distinctive characters, and the alveolar walls having partaken of the same fatty and granular change. The centre of the nodule is made up of a structureless, granular, or caseous mass, in which here and there only can any sign of the alveolar walls be made out. The chief part of the walls, similarly degenerated, has become fused with the alveolar contents. If the blood-vessels of a lung affected in the way we have described have been injected, it is found that the central part of these consolidations are free from injection, and that in the next zone the vessels are fewer than normally they should be, and no longer exhibit the well-known and characteristic appearance of pulmonary capillaries, but stop short in straight vessels. The caseous material in the centre of the lobule resists the action of dyes—as, indeed, this material always does wherever found. It should be noted that the caseation of the central zone is absent from some of the more recent nodules, which then exhibit the phenomena of catarrhal pneumonia, with an indication that part of the epithelial contents of the alveoli has commenced to undergo fatty degeneration.

ii. Croupous pneumonic consolidation.—In some of the consolidated patches the alveolar contents are occasionally found to exhibit an appearance which approaches that which is usually considered to be characteristic of croupous pneumonia; a considerable amount of fibrinous material binds them together. There is this difference, however, that the fibrin is in less amount than in ordinary croupous pneumonia, and that the number of epithelial cells and leucocytes is larger. This form of consolidation is therefore called by some by a special name,—caseous pneumonia, and occupies an intermediate position between the catarrhal and croupous forms. Alveoli affected with true croupous pneumonia are, however, sometimes to be met with.

Consolidated patches of any of the above forms

may coalesce to form larger patches until a whole lobe may be involved, and as their degeneration and softening proceed *pari passu* with their increase in size and number, so we may find collections of softened material like pus occupying the centres of some masses and entirely replacing others.

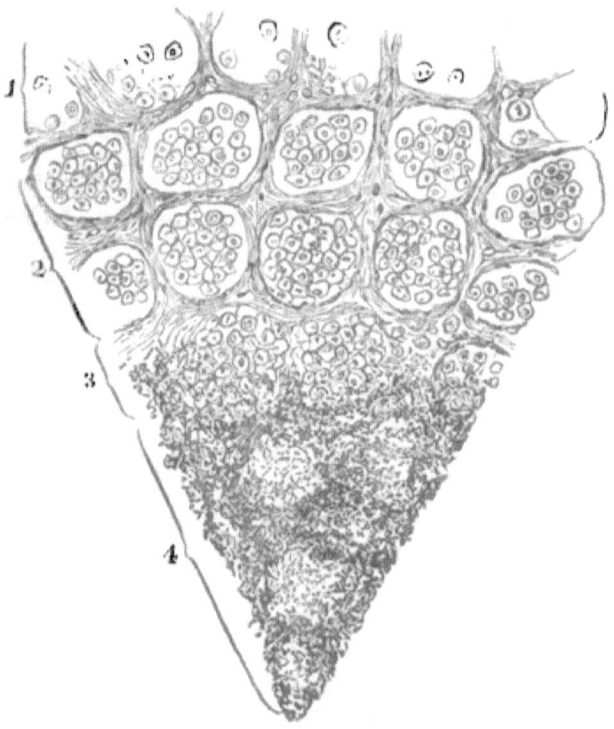

FIG. 37.—Segment of a rounded mass of consolidation from a case of chronic pneumonic phthisis. 1, The alveoli contain few loose epithelial cells; 2, the alveoli are full of epithelial cells and leucocytes; 3, the alveoli and contents undergoing solution; 4, caseous detritus.

These intra-alveolar inflammatory consolidations are by far the most common varieties of phthisical consolidations of the lungs.

iii. **Fibrous tissue.**—In all parts of the lung-tissue where the alveoli are stuffed full of inflammatory débris, the alveolar walls are almost certainly affected, and undergo more or less inflammation, with a multiplication of the connective-tissue cells and exudation of leucocytes. If the phthisical process be sufficiently chronic this may result in a development of fibrous tissue, but as a rule the alveolar walls thus inflamed take part in the degene-

Fig. 38.—A large miliary tubercle, formed of the coalescence of two or more tubercles, from a case of acute tuberculosis. The pulmonary vessels are injected. The alveoli are unaffected. Indications of the structure of the tubercle can be seen, but the power is too low for it to be made out distinctly. (Drawn from a photograph.)

ration of the alveolar contents. The adventitia of the blood-vessels is more or less inflamed and thickened. The inter-lobular connective tissue is generally increased in amount.

iv. **Bronchial ulceration and peri-bronchitis.**—In chronic phthisis, ulceration to a greater or less extent is found in the bronchi, whereby their mucous surface is made uneven and irregular, even if it is not entirely eroded. The connective tissue around them is also inflamed and in some cases greatly thickened. This is one of the ways by which the lesion spreads, as the alveoli in the neighbourhood of this peri-bronchitis are often secondarily affected.

v. True tubercle or grey granulation.—In addition to the consolidations above described, there may be found in the lung-tissue minute granular bodies which are called miliary tubercles. Occasionally by the coalescence of several tubercles the granules are

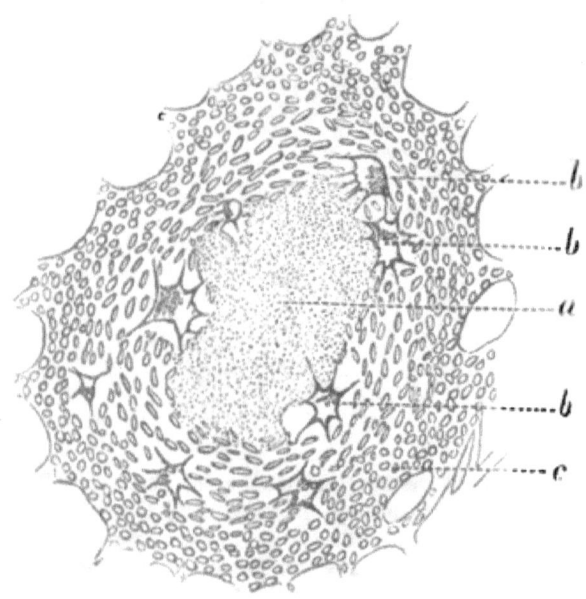

Fig. 39.—A smaller tubercle from the same case as the last figure under a higher power (Zeiss D, oc. 3). The zones are very distinct. *a*, Caseous centre; *b*, giant-cells; *c*, stroma of periphery arising from the processes of the giant-cells.

larger (Fig. 38), but they never approach the size of the intra-alveolar consolidations. They are grey in colour and hard. They are found in the neighbourhood of the caseous masses above described, in the mucous membrane of the bronchi, and elsewhere, but most typically in the pleura or sub-pleural tissue. Almost invariably they take their origin in the alveolar walls, or in the interalveolar connective tissue, or in some cases in that around the blood-vessels. It is possible that they are connected with the lymphatic tissue, as they appear to follow its distribution very closely. A tubercle then is probably the result of chronic inflammation of the connective tissue, and especially of the lymphatic tissue. It is essentially extra-alveolar to begin with, at any rate in the immense majority of cases. The growth, if it begins in the alveolar wall, encroaches upon the alveolus, forming a projection into it covered by the alveolar lining. From this primary projection others in various directions encroach upon the neighbouring alveoli, and in a short time the cells of the tubercle are so blended with the alveolar epithelium as to be indistinguishable from it. A tubercle on microscopical examin-

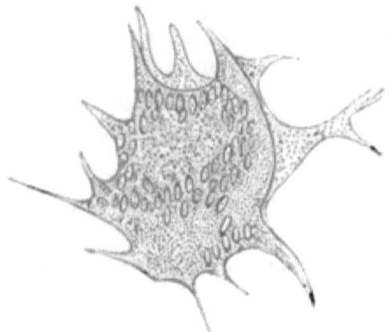

Fig. 40.—A giant-cell. (Drawn from a section of a very small young tubercle.)

ation at this stage looks like a closely aggregated mass of small cells like lymph-corpuscles.

In these masses may be detected large protoplasmic branching cells with many nuclei, the so-called *giant-, myeloid-,* or *mother-cells* (Fig. 39 and Fig. 40). These are said to be derived from the alveolar cells by the enlargement of one or by the coalescence of several of them, but more likely take their origin from the connective-tissue corpuscles.

*These giant-cells are **not** peculiar to, although they are very characteristic of, tubercles.* The processes of the giant-cells divide and anastomose, and thus form a meshwork at the periphery of the tubercle, and in the reticulum thus formed many of the nuclei of the processes lie. This reticulum is very fine and close, and forms a kind of sheath for the tubercle. The centre of a tubercle, when examined at a more advanced stage, is seen to have become granular and caseous, from degeneration. This change is due to the non-vascularity of the growth. Thus from within outwards a typical tubercle exhibits three zones, viz. a caseous degenerated centre, then a reticulum of small cells and giant-cells, and outside all a denser reticulum full of small cells.

Softened patches.—There is nothing at all characteristic in the composition or structure of the softened masses of caseous material except the presence within them of a large number of the micro-organisms already mentioned, viz. of the tubercle bacillus of Koch. These bacilli may be stained by means of fuchsine and other aniline dyes. They are minute, straight, or curved rods, about a quarter to half the diameter of the red blood-corpuscle in length, and one fifth to one sixth of their length in thickness, rounded at the ends and beaded, clear spaces alternating with their stained parts. Not motile. They are often arranged in chains, or collected into groups. Their propagation by spores is doubtful. In the tissues they are usually found in the caseous material, but have been discovered in the giant-cells, among the cellular contents of the alveoli, and in the newly-formed fibrous tissues of phthisical lesions. Outside of the body the bacilli can be grown on solidified blood-serum, or in Agar Agar mixture, but are very slow in growth, even at the most favorable temperature ($30°$—$40°$ C.). The micro-organisms may be destroyed by a high temperature, and by solutions of carbolic acid and of perchloride of mercury.

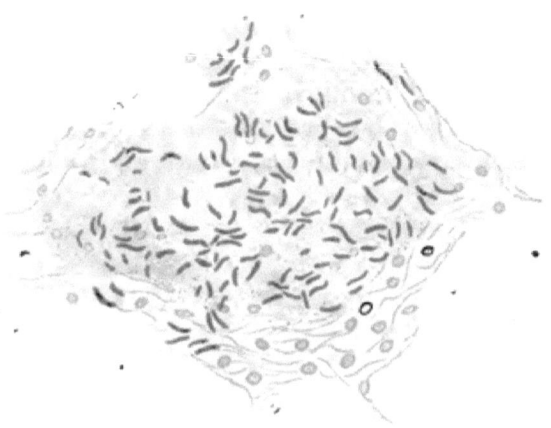

Fig. 41.—A nodule of interalveolar fibrous tissue. The fibrous tissue is broken down in the centre, and the débris is full of bacilli; the next stage would be a small cavity. From a case of tuberculosis arising at a late stage of chronic fibroid phthisis.

Walls of cavities.—The microscopical examination of the walls of mature pulmonary cavities shows them to be made up of two layers,—an inner layer, which is a more or less thick or pyogenic lining and externally a fibro-vascular coat. The lining can scarcely ever be looked upon as a secreting membrane, as it is rather by its own destruction from the inside that the purulent material has its origin. Sometimes, however, it appears to take on the function of a true pus-secreting membrane. The fibro-vascular layer is very full of blood-vessels internally, but outside, the fibrous tissue is more developed. As, however, the cavity becomes more and more quiescent, the fibrous portion increases at the expense of the vascular, and with it the irritability of the cavity and its secretion diminish. Small quiescent cavities with firm fibrous walls are very likely to be mistaken for dilated bronchi.

In recently formed cavities there is, as a rule, scarcely any attempt at the formation of a fibroid capsule. The inner surface is covered by the remains of the softened caseous tissue, so that if a stream of water be directed upon it the irregularities of the erosion are still further brought to the front. But it must be distinctly remembered that, as in some cases,

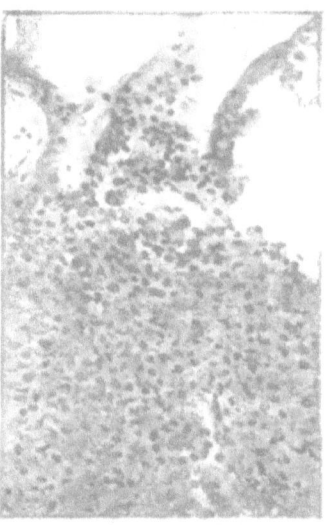

Fig. 42.—Section about half through a thickened pleura, and the neighbouring alveoli, to show the encroachment of the fibroid change on the pleura into the lung-tissue.

the changes in the alveolar walls of the boundary of an affected district go on at the same time as the changes in the centre of the district, the surrounding tissue forms a firm even if not a complete fibrous, investment of the newly forming cavity. It has been pointed out by careful observers, however, that often the necrotic change in the affected district follows so closely upon the inflammation which precedes it as to allow insufficient time for this fibrous reaction as it were to set in, and the result is that no line of demarcation is set up between the diseased and healthy tissue and great destruction is the result. This rapid destruction of lung-tissue, with the formation of ragged irregular cavities without fibrous capsule, one would expect to proceed from some special obstruction to the blood-vessels supplying the part.

vi. As regards the **fibroid changes** the amount of fibroid induration in a consolidated lung varies extremely, sometimes it forms the chief lesion. This is the case when the disease is traceable to the constant inhalation of dust or of irritating particles, and constitutes the conditions known as *anthracosis pulmonum* and *siderosis pulmonum*, which are produced by the inhalation of coal and iron dust. As the result of this the appearances of the lung post mortem are such as have been described under the head of interstitial pneumonia or fibroid induration; with it the formation of bronchiectases is closely associated. From this condition to that of cases of acute miliary tuberculosis, in which there is hardly any fibroid change, there are gradations in which the fibroid material appears in diminishing amount. Healing and partial obliteration of cavities, demarcations of diseased from healthy tissue, when they occur, or encapsulation of calcareous masses, take place by means of this tissue. It arises in the alveolar wall and in the interalveolar and interlobular tissue, or in the peri-bronchial and peri-arterial tissue. It is as a rule well supplied with blood-vessels. Sometimes, however, it appears that the peri-arterial tissue goes on to more or less complete obliteration of a vessel supplying a tract of the newly formed fibroid tissue, which therefore undergoes caseation. The tract is usually round or sinuous. The centre may soften and a cavity with ragged edges which tends to increase in size is the result (Fig. 41).

Thus it seems that cavities may appear bounded

by fibrous tissue may arise not only from the growth of that material in the neighbourhood of interalveolar consolidations but also from their formation in the midst of fibroid tissue of recent and inflammatory origin.

The relation in amount between the various factors which together make up the post-mortem appearances in a chronically diseased lung must, of course, vary very much. No doubt these variations are the true cause of the immense variation in the symptoms and course of phthisis in different individuals.

The post-mortem appearances in a case of what is usually called **acute miliary tuberculosis** of the lung must now be mentioned. In this affection the lungs are almost uniformly studded from apex to base with miliary tubercles, which are rounded, grey, and translucent, and the pleura is also granulated with similar minute growths, which are found elsewhere in the serous and mucous membranes. The pulmonary tissue is injected and often more or less infiltrated with blood. There is besides considerable hypostatic congestion of the bases of the lungs. Tubercles are usually found in the larynx, arachnoid membrane, and elsewhere.

The tubercles in this form of the disease, present in a more or less typical way the minute structure of a miliary tubercle which has been described above. If the affection, however, has been very acute, the alveoli in the neighbourhood of the tubercle may exhibit signs either of (a) croupous, (b) catarrhal, or (c) insterstitial pneumonia, or of all three.

Pathology.—The pathology of phthisis, upon which so much depends, is very complicated, and is still in considerable confusion. Up to a few years ago there were three chief views upon the subject, the first, which was taught in great part by Laennec, was that tubercle is a neoplasm or non-inflammatory new growth, one of the characteristics of which is a proneness to undergo caseation, a degeneration which does not attack other new growths; that in phthisis the lesions all exhibit a proneness to undergo caseation, and that therefore all phthisis is tubercular. In short, Laennec considered phthisis to be a constitutional disease manifested by the evolution of pulmonary tubercles. Now, although this was Laennec's doctrine, it is only fair to remember that he recognised the naked-eye or main division of tubercles into the minute or isolated and the infiltrated. His views as to the nature of phthisis were supported by clinical as well as pathological observers, and were the received opinion upon the subject for many years. Two forms were recognised, but were considered to be essentially similar in origin, differing only in amount. After about twenty years, however, it was shown, firstly, that other masses besides tubercle tend to caseate, secondly, that many of the supposed tubercles of the lungs were the collected products of inflammation and were not tubercle, and that, thirdly, giant-cells, which had been considered to be characteristic of tubercle, were found in many other places besides. These researches paved the way for the second view of phthisis, which was most ably advocated by Niemeyer and is generally associated with his name. According to this, phthisis is seldom tubercular, if the term tubercle be restricted to the isolated or miliary tubercle, as Virchow had suggested. The results of microscopical examination had demonstrated that the miliary tubercle differed from the so-called infiltrated tubercle in structure, and that the latter, too, did not manifest always the same histological appearance. Niemeyer stated that the most common lesion of phthisis discovered postmortem is inflammatory consolidation of the tissue of the lungs in smaller or larger patches, *i. e.* caseous pneumonia, and that the most common cause of the consolidation is chronic catarrhal or bronchopneumonia of the apex of the lung. Next in point of frequency, as a cause of consolidation, he counted acute catarrhal pneumonia, and finally he considered that occasionally the lesion might be of the croupous form. He also separated into a distinct form, interstitial pneumonia or fibroid phthisis, as it is often called. The way in which these lesions arose, and in which they produced phthisis, was, according to his theory, as follows:—A strumous person is attacked with bronchial catarrh, which extends into the pulmonary alveoli. The products of the alveolar inflammation, both epithelial and fibrinous, instead of being got rid of in due course by expectoration, remain behind and undergo caseation, producing thereby the pulmonary consolidation. This consolidation may undergo any of the following changes: (*a*) The caseous material may, after a longer or shorter time, under favorable conditions, be *got rid of by absorption or expectoration*, liquefaction having previously taken place. If this occurs speedily the lung-tissue is little damaged; (*b*) It may undergo *calcification*, doing little permanent injury; (*c*) It may *soften and break down*, involving the lung-tissue in its destruction. A *cavity* in the lung-tissue might thus result. This method of cavity-formation Niemeyer looked upon as the usual one; (*d*) A secondary and most disastrous effect of this liquefaction may be the *inoculation of the neighbouring lymphatic or blood-vessels* with some of the softened débris, whereby the minute vessels in the neighbourhood or throughout the system generally become blocked by minute emboli.

These emboli serve as minute foci of inflammatory action, with the result of producing *true miliary tubercles*. If the vessels are only locally affected, then the tubercles are only local, causing a local tuberculosis confined to one lung or parts of it, but if the vessels are involved generally the affection is general, and an acute general tuberculosis is the result. Niemeyer's dictum which described this last disastrous result was that *the greatest danger to most phthisical patients is the development of tubercle.* According to this theory tubercle is always secondary to a softened caseous mass, and if such a caseous mass be not found in the lung it must be looked for elsewhere. Thus it may be found in the intestine, in cases of scrofulous ulceration of the intestine, in the pleura, in the pericardium, or in a strumous knee-joint, as the result of preceding inflammation.

The **third** theory as to the pathology of phthisis enunciated by Rindfleisch before the discovery of the specific micro-organism was to this effect: That when strumous people become affected with catarrh of the apices or other parts of the lungs, in the catarrhal secretion is contained the tubercular poison, which may inoculate the edges and corners of the narrowest portions of the bronchi, and that as a result of this a tuberculous infiltration arises of all the angles and projections situated at the points where the smallest

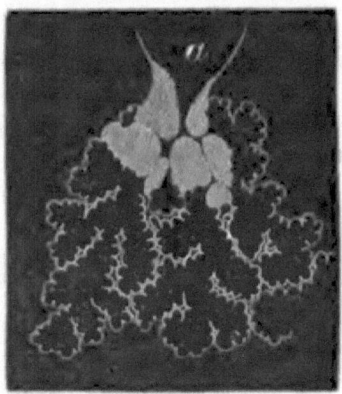

FIG. 43.—Diagram of primary tubercle granulum of the lungs. Infiltration of all the connective edges at the point of entrance of a bronchiole. *a.* An enlarged bronchiole. (Rindfleisch.)

bronchioles become continuous with the alveoli of the lung (Fig. 43.) That in this way circumscribed white nodules are formed, corresponding to the isolated tubercles or tubercle granules of Laennec, in other words a cellular infiltration of the pulmonary connective tissue occurs composed of small bloodless nodules incapable of suppuration, resorption, or organisation, but only of degeneration. That later on in the affection, the mucous membrane of the small bronchi becomes ulcerated, and that the tubercular growth in consequence arises in the walls of the ulcers, and then infects the lymphatics in the peribronchial connective tissue. These lesions, tubercular bronchitis and peri-bronchitis, gradually extend. That secondary to this, the alveoli of the lungs become the subject of caseous pneumonia, which considerably enlarges the diseased area, joining together into compact masses smaller areas of tubercular inflammation.

The discovery of the tubercle bacillus in the lesions of every recognised variety of phthisis has led to an alteration in the views as to the pathology of the disease. Inoculations of pure cultivations of the bacillus have been shown to set up general tuberculosis in certain animals. These experiments have proved the specific origin of miliary tuberculosis, which had been strongly suspected for

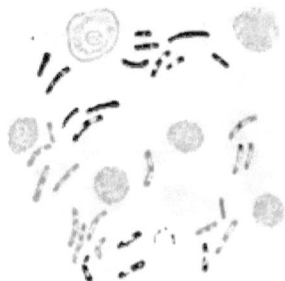

FIG 44.—Tubercle bacilli in the sputum.

many years. Many pathologists go a step further. They consider that it has also been proved that all the lesions of phthisis are due to the bacillus, and that the varieties in the lesions depend upon the amount of the specific organisms introduced, and the intensity of the inflammation set up by them. Thus if the irritation be great, alveolar inflammation, chiefly consolidation in larger or smaller patches, results; if it be inconsiderable the inflammation is for a long time confined to small areas, and particularly to the alveolar walls, the extra-alveolar tissue, or the connective tissue elsewhere. Further, that miliary tubercles proper may or may not arise under such circumstances. These views are very possibly correct. Our own impression, however, upon the subject is that isolated miliary tubercles must always be considered to be secondary, and that they are due to the local infection of the lymphatic or blood-vessels by the tubercle bacillus, just in the same way as general miliary tuberculosis is produced by the general infection of the lymphatic or blood-vessels by the same means.

It is not by any means clear that the alveolar inflammations are always set up by the bacillus, although this is possible. If they are so induced, the lesions are primary and are to be distinguished from miliary tubercle. It is, however, *quite possible that the initial inflammatory lesion is non-specific*. Thus, suppose a case of catarrhal pneumonia to arise in a strumous person, that the inflammatory products within the alveoli undergo caseation, and that the bacillus is introduced into the caseous mass thus produced, the following results may be expected: (1) that the softened material may be expectorated, and by means of fibroid tissue which has been developed in the neighbourhood of the mass, the lung substance may be able to resist the further action of the bacillus; (2) that by the irritation of their presence the micro-organisms may cause the inflammation to spread into neighbouring alveoli, with the production of further caseous material into which they further extend their operation; (3) that from the breaking-down caseous mass as a focus, they may spread into the neighbouring lymphatics and produce tubercles, which may degenerate and blend with the primary lesion and so increase its size; or (4) that they may infect the lymphatic or blood-vessels generally, and so produce general tuberculosis.

This view of the operation of the specific virus would agree with Niemeyer's theory, and would not be inconsistent with the difference in the lesions of the disease as examined under the microscope.

It will also explain the cases of phthisis which may occur (as is generally admitted) as sequelæ of croupous pneumonia or catarrhal pneumonia, which cannot have arisen in the first case from the presence of the tubercle bacillus.

The fact that the majority of those who breathe the air, which has been proved to be very full of the bacilli, in chest hospitals, escape the disease, has been explained by the suggestion that the bacillus cannot enter, or if it enter, that it cannot live unless a predisposition to the disease exist. A more likely explanation appears to be that the bacilli do enter the lungs of all who breathe an atmosphere contaminated with them. They, however, grow slowly and require a specific kind of nourishment. The latter they are unable to obtain in a healthy lung, but that in a lung with patches of caseous pneumonia in it, what they require exists, and if introduced into a caseous mass or into unhealthy alveoli, unless they are speedily got rid of, they may multiply and produce the effects before described.

Symptoms and Course.—The onset of phthisis is as a rule insidious. A patient may gradually lose health without showing any symptoms of a chest complaint, until for some reason he consults a medical man, who discovers on examination of the chest the evidence of distinct consolidation at the apex. In another case attention may be attracted to a patient's condition first of all by an attack of blood-spitting or by some febrile symptom. In a third case symptoms which ought to give warning of the disease, such as cough or slight expectoration, are overlooked because the general health is apparently unaffected, till patient is at last warned of the seriousness of his complaint by increasing weakness and loss of flesh, and possibly by sleep-sweats. Again, phthisis may develop in a patient subject for years to winter cough, and he is warned of his new danger when the cough and dyspnœa do not abate in the summer as usual. A considerable number of consumptives date the commencement of their illness from "breaking a blood-vessel," others from a severe cold caught in a fog, a snowstorm or the like, upon a definite day. The symptoms as well as the course of the disease very much depend upon the presence or absence of complications either in the larynx, intestines, or elsewhere.

Considering the immense variety in the lesions of phthisis, the different circumstances under which it occurs, its long duration in many cases, and the many complications to which it may give rise, it is extremely difficult to arrange the symptoms of the disease in anything like order. The diagnosis of consumption in the lung indeed depends to a considerable extent upon the physical examination of the chest, and symptoms are chiefly valuable in a majority of cases when considered in conjunction with physical signs. We will, however, consider some of the chief or most common of them in order.

Cough.—Cough is almost an invariable symptom of consumption. It is usually one of the first symptoms, if not the very first. It varies in amount, from the merest "hack" to a constant wearying trouble, sometimes worse in the morning or at night, and increased on exertion. It may be for some time a winter cough, being absent in the summer, and, indeed, the presence of a cough unabated all the year through, is justly considered by the laity to be a serious matter. It does not appear to change so much with change in the weather as bronchitic and other simply catarrhal coughs. It may, however, be a slight cough, which scarcely does more than annoy, but continues for a very long time. We have known patients in an early stage of consumption say that they have had the cough for four, five, or six years, but have never troubled about it. It is seldom paroxysmal.

Spitting.—At first there may be only a little glairy catarrhal mucus. As the affection advances, the sputum, however, becomes more profuse and muco-purulent. In later stages it is purulent, thick, nummular and greenish, airless, although coated as it were by mucus containing its characteristic air-bubbles, sinking in water, but kept near the surface by the adhering mucus. The appearances of the sputum are not characteristic to the naked eye. If caseous masses or minute pieces of lung-tissue or elastic fibres be made out under the microscope, some destructive disease of the lung may be diagnosed, and if a fair quantity of tubercle bacilli be discovered on several separate examinations in the sputum by means of appropriate staining (p. 58), consumption of some kind is to be inferred. The non-discovery, however, we do not believe to indicate its absence, for two reasons, firstly, because there are non-bacillary stages of the disease, and secondly, because the bacilli may not have been detected. Not everyone can stain tubercle bacilli, and even the most skilful do not always succeed.

Shortness of breath.—This symptom, which may not be very marked at first, gradually increases. It means, of course, that the respirations are more rapid, whereas their depth is diminished. This passes in the later stages to very considerable dyspnœa, so that the slightest movement is attended with an extraordinary rapidity of breathing. The dyspnœa is partly due to diminution of the breathing surface in the lungs, partly to pain, and partly to fever.

Hectic fever.—The fever may be very slight at first, but very early it is found that the evening temperature is one or two degrees higher than natural, even if the day temperature is only just above normal. In some cases more than others the temperature is raised so as to be a marked feature. In the pneumonic form this is particularly the case. In acute phthisis we have known it to vary constantly from 101° F. to 104° F. With this raised temperature the patients suffer from **night sweats and sleep sweats** of the most exhausting nature, sometimes confined to one part of the body or to the extremities, sometimes general. The sweating must be considered a distinct act of secretion by the sweat-glands, and not as merely due to the dilatation of the cutaneous vessels. It is supposed to arise from the increasing venosity of the blood during sleep stimulating the sweat-centre or centres. The patients wake up in a most profuse sweat, the amount of which is sufficient to saturate their bedclothes. Emaciation is sometimes exceedingly rapid; the muscles waste and are exceedingly irritable, so that when slightly stimulated by a gentle tap with the finger contract and flicker for some time (myoidema); the subcutaneous fat is rapidly absorbed, leaving all the bones marked out. This is particularly noticed in the face, the nose is thin and prominent, the eyes sunken, the jaws sharp, and the temporal fossa hollowed out. The hair becomes thin, the gums spongy and red, the appetite capricious, with more or less digestive troubles; the fingers are clubbed.

One of the features of the disease is the

extraordinary **hopefulness** of the patients, and their belief in their ultimate recovery, even although to the bystander it is evident that death cannot be long delayed.

Hæmoptysis.—The relationship between hæmoptysis and phthisis is not simple. In no less than two thirds or even three fourths of the cases hæmoptysis occurs at some period or another. Many consumptives say that their illness began by their "bursting a blood-vessel." It is often a symptom of the advanced disease, and not infrequently is absent until quite late in the affection. At other times it occurs off and on during its whole course, and often it is the first symptom, and possibly the cause of phthisis, as has been already pointed out. The quantity of the blood also varies very much, from mere streaks to a gush, which may be fatal. It is as often moderate as profuse. The blood is either pure and very fluid, or it may be mixed with phlegm or food if vomiting occurs simultaneously. When much mixed with phlegm the pneumonic form of phthisis is to be expected. The hæmorrhage may be a mere capillary oozing, but it may also arise from bursting of a minute aneurysm of a small branch of the pulmonary artery.

Pain in the chest and shoulders.—Pain in the chest is seldom absent throughout the whole course of the phthisis. Nothing is more common than for phthisical patients to complain again and again of some thoracic pain as the disease advances. The pain is pleuritic, in the great majority of cases, and indicates that the lesion has extended to the alveoli. It occurs more often in the pneumonic than in the pure tubercular form of phthisis.

Other symptoms:

Hoarseness of voice or aphonia, which is either

due to tubercular ulceration or to chronic catarrh of the larynx. There seems to be considerable evidence in favour of the belief that tubercles may and often do occur in the larynx before they are developed in the lung.

Diarrhœa.—The diarrhœa due to tubercular ulceration of the intestine is generally a late symptom, and is comparatively a rare one.

Albuminuria.—Albuminuria is certainly not common in any form of phthisis, but the kidneys, ureters and bladder may often be affected by tubercular disease, and the first by interstitial nephritis.

Œdema of the feet and legs occurs chiefly in the advanced disease from pulmonary obstruction. Occasionally there is thrombosis of some vein which causes a local œdema.

Fistula in ano is not an unfrequent complication. It appears, according to our experience, most often in men.

Cerebral symptoms may occur when the tuberculosis has become general from tubercular meningitis. Tubercles may sometimes be detected in the choroid under such circumstances, or optic neuritis.

Physical Signs:
Vary according to the stages of the affection, which are recognised, viz.: i. **Consolidation.** ii. **Breaking down.** iii. **Loss of substance or cavity.**

Inspection.—The phthisical predisposition is often exhibited. The patient presents one of the various types of the so-called strumous habit. Bones are slender; skin thin; cheeks red; sclerotics bluish; subcutaneous connective tissue contains little fat; muscles are ill developed; neck is long; intercostal spaces are broad; the sternum is at an acute angle with the ribs; the entire chest is flatter, narrower, and longer than in health. Scapulæ stand out like wings. In the chest there is more or less fattening in various situations, especially in the supra- or infra-clavicular regions, according to the position and extent of the disease, but it is seldom met with until the third stage. Impaired movement is noticed in all stages. The heart's apex-beat may be diffused and displaced.

Palpation.—The vocal fremitus is increased over the affected portions of the lungs in all stages.

Percussion.—Dulness over the consolidated or breaking-down parts. There is usually dulness over a cavity unless it be deep and covered by healthy lung, when the percussion is unaffected; when the cavity is very large, there is hyper-resonance. A cracked-pot sound may be detected in percussing over a cavity if superficial and communicating with a bronchus; or there may be amphoric resonance.

Auscultation.—Feeble inspiration above or below one clavicle is one of the earliest signs of the affection, and this is soon accompanied by harsh or jerky and prolonged expiration. Crepitations may also be present very early with more or less sibilus and rhonchus. Later on bronchial breathing and bronchophony with, or followed in the next stage by, subcrepitant râles. When a cavity is established there are gurgling râles, cavernous or amphoric breathing. Metallic tinkling over a large empty cavity, with smooth, regular, and firm walls.

Varieties and course.—Even those who acknowledge but one kind of phthisis, viz. the tubercular, will admit that clinically there are several types of the affection as regards onset and course.

The division of the cases into acute and chronic is one of considerable convenience, although it must be recollected that phthisis, apart from general miliary tuberculosis, is never an acute disease in the ordinary meaning of the term, and certainly lasts some months. There is little characteristic in the onset and course of disease which is diagnostic of any special variety. The *chief distinction between the acute and chronic forms* is that in the latter the course of the disease is nearly always every now and then arrested by periods of quiescence and even of improvement and healing, whereas in the former the course is uninterrupted or continued, and the patient goes from bad to worse, and dies in a few months' time. The cause of this difference is exceedingly obscure. **Acute miliary tuberculosis** may arise independently of lung disease, and is accompanied by the constitutional symptoms of high fever, not unlike those of typhus, or those of meningitis together with the local signs of bronchial catarrh. It may terminate in three or four weeks. It may also arise in the course of pulmonary phthisis, being the last stage of very chronic cases. As regards the **pneumonic** form of the disease, which may be either acute or chronic, the extent of the lesion determines its course.

One cannot diagnose from the physical signs or symptoms the kind of inflammation we have to deal with; it may be catarrhal, cheesy, or croupous. The most *rapid cases (phthisis florida or galloping consumption) are those in which large tracts, either groups of lobules or even lobes, are suddenly attacked* by pneumonia. This is usually primary, but may occur in the course of a chronic case, and cause its speedy conclusion. As regards **fibroid phthisis or interstitial pneumonia**, this is a chronic affection essentially, and is accompanied by physical signs much more marked than its clinical symptoms. Finally, **syphilitic phthisis** is an affection seldom recognised during life but is chronic in its nature.

Terminations.—i. In partial or complete recovery, not so rare as was formerly supposed.

ii. In death, after a varying time from a few months to four or five years, or even more.

Generally death is very gradual, and due to a slow form of asphyxia. It may be due to **diarrhœa** or **other** complications.

Aphthæ on the tongue are common at the last.

The *complications* which are likely to cause death are as follows: *Pneumothorax. Hæmorrhage*, from giving **way** of some large **vessel.** *Thrombosis* in the **extremities.** *Intestinal ulceration. Tubercular meningitis.* In addition to these, **other** complications make the prognosis bad, for example, *pleurisy* or *lobar pneumonia* added to the original affection. *Amyloid disease of the kidneys. Laryngitis.*

Duration and prognosis.—Both the prognosis and the duration of phthisis depend much upon the variety of the affection with which we have to deal, and before an **answer can be given** to the question, **Is there any chance of recovery?** an exact diagnosis must be attempted.

Treatment.—Although it by no means follows that all the children of tubercular parents become phthisical, yet there are sufficient grounds for carefully looking after the health of the members of a family in which there is a strong history of phthisis, and especially of an individual member who may have shown signs of defective nutrition. The means adopted in such cases should also be **adopted** in any **other** cases in which an acquired tendency to consumption is discovered. These means are summed up in the following rules:

- The **diet** should be nutritious and easily digested, and should include a considerable amount of milk. As far as our experience goes, alcoholic stimulants in moderation are of **use** to adults, such as light **bitter** ale, very light dry sherry, or the like; they should be taken at dinner time.
- **Cold baths.**—With careful shampooing, especially of the chest. If the patient is unable to use absolutely **cold** water, then water, tepid or as **cold as** possible, may be substituted.

Fresh air.—There is no doubt that an outdoor life is the best for patients with a tendency to what is called a weak chest; the country better than the town. At any rate, delicate persons should not be allowed to remain shut up in close, ill-ventilated shops or workrooms. This necessitates careful consideration in the choice of occupation for boys and girls of delicate families.

Exercise.—A certain but not too great amount of daily muscular exercise, walking, riding, cycling, or what not, should be enjoined. Drilling also is of much use to boys and girls with narrow chests, and moderate athletic exercises. Out-of-door games, such as cricket and tennis, if indulged in with reason, cannot be too strongly recommended as likely to develop a condition of constitution which resists phthisis.

Warm clothing.—Patients should be recommended to wear flannel next their skin.

If the delicacy of constitution be already marked, but without the appearance of chest complaint, then, in addition, residence at the seaside, or a sea voyage in a sailing ship, or a prolonged stay at some chalybeate spring may be of use. Cod-liver oil should be administered.

The following advice cannot be too strongly insisted upon:

Marriages of phthisical persons are to be condemned.

Trades necessitating the breathing in of irritating particles, or injurious smells, or cramping the respiratory muscles are to be avoided.

Prolonged lactation, over-study, late hours, and over-exertion of mind and body, should also be avoided.

Catarrh must be treated as though a serious disease.

When the diagnosis of phthisis has been arrived

at, the treatment depends much upon whether the disease is—i, pneumonic; ii, fibroid, or, iii, tubercular proper; or if it be thought that all phthisis is bacillary, whether the changes are localised to the primary seat, or whether they have spread and become more general. In this section, for the sake of convenience, we will assume the existence of three forms of the disease.

    i. **If pneumonic.** If there is fever, Niemeyer strongly recommends keeping the patient in bed for a time, and treating him with poultices, blisters, slop diet, antipyretics, and, indeed, as though he were suffering from an ordinary attack of pneumonia. It is undoubtedly true that rest in bed, quiet, good diet and counter-irritation to the chest often work wonders in patients, and that under these conditions, as, for example, in a consumption hospital, the patients increase in weight and strength, and that their coughs and sweats considerably diminish, as well as the physical signs of chest disease. After such improvement the patient should, if possible, be sent for change of air to one or other of those places to which we shall again allude, which have been found favorable as regards climate and position for phthisical patients. If their circumstances do not allow of prolonged idleness, then the question of a change of occupation and of permanent change of habitation should be considered.

    In advanced cases, these questions become still more pressing, but it should be recollected that for a very long time attention to hygiene, good food, proper exercise and fresh air greatly benefit the patient and prolong his life.

    Finally, treatment of symptoms becomes the only indication.

    ii. **In fibroid phthisis**, removal from all causes of irritation, and treatment by change of occupation and climate, should be advised as in the pneumonic form.

iii. In tubercular phthisis proper, or those cases in which the infection from having been local has become general, or at any rate distributed throughout the lungs, and in the rare cases of primary miliary tuberculosis, it is almost useless to expect much good to result from expatriating the patient. Some alleviations of symptoms may, however, be looked for by treatment.

In any case of phthisis it may be necessary to treat symptoms, and it will be as well to summarise some of the usual methods which are found useful.

Symptomatic Treatment.

Cough. — Counter-irritation, especially in the form of small flying blisters to the chest; opium or morphia in small doses do good. Tonics, especially strychnine, also appear to allay cough. Where there is much attendant bronchial catarrh, expectorants are useful.

Fever.—Quinine (gr. iij—v ter die), and stimulant diet, and wine are what are usually employed. Antipyrin (gr. v—x) has also proved of value.

Night sweats. — Sponging with vinegar and water or with plain water is grateful to the patient. Dilute acids, especially Acid. Sulphuric. Aromat. ♏xx—♏xxx ex aqua, at sleep time often do good. Atropine gr. $\frac{1}{100}$—$\frac{1}{50}$ in formâ pilulæ, is often efficacious, but sometimes produces its physiological effects, and so cannot be employed, in which cases picrotoxine is often of use. Oxide of zinc (gr. ij—iv) is not so much used for checking night sweats as it was, but is certainly of use in some cases. Salicylate of sodium (gr. xx ter die) is also recommended.

Hæmoptysis.—Ergot in some form appears to exercise the most influence in subduing blood-spitting (Ext. Ergot. Liq. ʒj or more, 6tis horis, or ergotin injected hypodermically). **Hazeline** is also a hæmostatic which may be employed, and such remedies as gallic acid. But with the drugs absolute rest and quiet, cold and slop diet, purgatives, and ice to suck must be combined. A blister or mustard plaster is often employed. Digitalis is not as a rule of much use in these cases.

Pain.—Generally some counter-irritant subdues the pain in the chest, such as a **mustard poultice, iodine liniment** or a **blister.** The pain of a thrombosed limb is relieved by liniment of belladonna or of opium.

Insomnia.—Phthisical patients get into the habit of taking a considerable amount of opium preparations for cough and sleeplessness. **Chlorodyne** (♏xx or more) at night, **chloral** (Syrup. ʒj) may be employed with advantage. **Urethane** is well spoken of by our resident medical officers at the Victoria Park Hospital as often producing tranquil and healthy sleep.

Vomiting.—A frequent cause of distress to phthisical patients is often relieved by an effervescing mixture made with infusion of calumba instead of water.

Diarrhœa.—If the ordinary astringents used in diarrhœa fail, **bismuth** is often efficacious, either the subnitrate (gr. v to x) alone, or combined with conium or hyoscyamus. **Sulphate of copper** (gr. $\frac{1}{4}$) and extract of opium (gr. $\frac{1}{4}$) combined in the form of a pill, seem sometimes to stop the most severe diarrhœic attack.

Hoarseness and pain in the throat are some of the most difficult symptoms to relieve. Inhalations of conium and the like are seldom much good. Local applications to the pharynx of glycerine of morphia, or to the larynx in the form of powders, are sometimes of use. **Cocaine** lozenges are also recommended, and the solution of cocaine (10 per cent.) applied locally.

Wastings.—In nearly all cases cod-liver oil may be tried. A patient, who is quite unable to take a dark oil, will often take easily another variety more clarified. In place of oil, if quite out of the question, bacon fat or cream may be tried.

Malt extract or **maltine** is beneficial, and may be taken in many cases by patients to whom oil is repugnant; or maltine and iron.

Dietetic Treatment.—The feeding of patients who are suffering from phthisis is of the greatest importance, and various of the so-called "cures" depend, to a considerable extent, upon some special method of dietary. As before mentioned, milk should enter largely into the diet scale, and if not borne undiluted, should be mixed with soda water, Apollinaris, or some other mineral water. Sometimes, when cow's milk cannot be taken, the milk of other animals, the donkey or goat, is for some reason or another taken with relish and digested; and this notwithstanding the peculiar twang noticed by one unaccustomed to take goat's milk. Even in London, goats appear to flourish; and some time ago we had a patient living in one of the suburbs who supplied himself with plenty of milk from his own goats and became quite stout and strong upon it. The cost of keeping the goats, he told me, was very small. When milk cannot be borne, koumiss should be tried. It is obtainable in London, being supplied by the Aylesbury and other Dairy Companies in three strengths.

Whey and **buttermilk** are strongly believed in by some.

Whether **cod-liver oil** is to be looked upon simply

as a food or not is a matter of opinion, but at any rate such fatty foods appear to be distinctly indicated.

Dr Dobell's **pancreatic emulsion** is often tolerated when the fish oil cannot be taken, but we doubt strongly the theory upon which its administration was recommended.

Alcohol in some form is strongly advised. Nearly all those who have had much special experience of phthisis agree in this; for example, Dr H. Weber says, "In some cases as much as a bottle, and even three pints, of moderately strong wine, or ten to twelve ounces of cognac or whisky are taken in twenty-four hours with advantage; in others scarcely one sixth of this amount, and again in others alcohol must be altogether avoided."

Frequent small meals during the day are to be recommended.

As regards the drugs specially indicated in phthisis, unless there is a tendency to hæmoptysis, iron is the chief; it may be given as the Tinct. Perchlorid., Tinct. Acetatis, Ferri et Ammon. Cit., Ferri et Quin. Cit., and in other forms. Arsenic, too, is often useful. Strychnine and the hypophosphites.

Climatic Treatment.—The main objects to be attained by recommending to a consumptive patient change of air are, firstly, that he may escape the dampness, mist, fog, and want of sunshine from which we are too apt to suffer, at any rate during the winter, in the majority of places in our own country; secondly, that he may obtain greater opportunities of being in the sunshine in the open air without exposing himself to the chance of catching cold; thirdly, that the air he breathes may be pure and aseptic; and fourthly, that he have sufficient change of scene or absence of worry as not to feel the withdrawal from his usual avocations. How are these indications to be met?

i. **In the case of a wealthy patient who can give up his whole time to his own cure or relief.**
According to the nature of his case, his disposition, and other personal circumstances, he may be induced to try—

(a) **A sea voyage,** or what is called by authorities the tonic marine climate, because of the breeziness, purity, and marine quality of the air; the voyage being to Australia, New Zealand, the Cape, the West Indies, or elsewhere, according to circumstances. A sailing ship is to be preferred to a steamer.

(b) **The high altitudes,** which are tonic and stimulant because of the rarefaction of the air, are pure and aseptic—but not so uniformly tonic as (a). Of these climates, Davos Platz, St Moritz, Wiesen, in Switzerland, and various less known places in the Andes, are specially recommended.

(c) **The southern health resorts,** including Algiers, Tangier, and Northern Africa, the islands in the Mediterranean, and the French and Italian Riviera, *e.g.* Hyères, Nice, Cannes, Mentone, Bordighera, and San Remo, &c.—Madeira, Teneriffe, Canary Islands, &c. These places have warm climates, varying in degree of heat, wind, and moisture of the atmosphere, some being dry marine resorts and desert sanatoria, partly tonic and stimulant, and others being moist marine climates, *e.g.* Madeira, which are relaxing and sedative. The patient may pass from one resort to another, according to the season of the year and other circumstances.

(d) **The home health resorts** furnish for the autumn and early winter many places suitable for the consumptive, *e.g.* Ventnor, Bournemouth, Sidmouth, Dawlish, Torquay, the South Cornwall coast to Penzance, Ilfracombe, Hastings, &c., and near London the pine woods of

Hampshire and Surrey, *e. g.* from Leith Hill and Farnborough to Ascot.

Our own opinion is strongly against laying down arbitrary rules **as to the exact** spot a phthisical patient **should seek**, according to the exact pathological condition made out in the chest *alone*; many other things must be considered: **age**, sex, duration of illness, disposition, general **health, and the like.** At the same time one ought to take notice that those **who know** most about it think **that** the following classes of patients should *not* be **sent to** the high altitude health **resorts, viz.: (1) Those** with a weak circulation, either due to organic heart disease or to other causes; (2) **the old,** especially those in whom senile changes **are distinct;** (3) the gouty, rheumatic, and those suffering with nervous diseases, hysteria, **or marked** dyspepsia; (4) those who **suffer** from **nervous and** circulatory excitability. All other **cases,** except where the disease is very **advanced, as a** rule do well at Davos and similar places, and are not only relieved, but in over **10** per cent., so the doctors say, are actually cured there. The Riviera, Algiers, Madeira, and other such places **can** boast with justice of their cures; and in spite of the extreme reaction in the medical profession in favour of cold and high climates **in** the treatment of consumption, many invalids will **always** prefer the **less** tonic and stimulating **or** even relaxing climates before mentioned, and, moreover, will often do better there than in the mountains.

ii. **In the case of patients who must work for their living** it becomes the duty of the physician to recommend that they, at any rate in the earlier **stages** of the complaint, should after a longer or **shorter interval** leave this country permanently for a climate more suitable, in which they can do some work even if not of the kind to which they have been accustomed; work which necessitates a more out-of-door life being the best. They may go **to Australia,**

New Zealand, the Cape of Good Hope, California, and the like, or if it be practicable they may resort to one of the sanitoria above mentioned, and take up their permanent abode, and look out for some light work there.

iii. **In the case of patients who cannot leave this country.** If they can afford it, a permanent abode in one of the above healthy seaside places may be tried, or others which might be mentioned. Any change from a large town into the real country, unless into damp, low-lying districts, is better than none.

Failing all these alternatives, the **treatment of consumptives during the winter in hospital, with a change now and again to a convalescent home, will frequently prolong life for many years.**

The treatment of certain stages of the disease by means of **antiseptic inhalations** is one which from our experience can be strongly recommended. About half a drachm of one of the following mixtures is dropped upon the sponge of a simple respirator, and the patient is directed to breathe through the respirator for an hour, three or four times a day. Under this treatment it often happens that the cough and expectoration diminish, the appetite is increased, the patient sleeps better and gains in weight. The inhalations are arranged so that if one is found too irritating to the pulmonary mucous membrane, another can be substituted. The formulæ are as follows: Olei Eucalypti, Alcohol Ethylici, āā partes æquales. —Olei Pini Sylvestris, Alcohol Ethylici, āā partes æquales.—Thymol, Camphoræ, āā 2 grms., Acidi Carbolici pur. 2 cc., Spiritus Rectificati 20 cc.—Olei Eucalypti 10 cc., Chloroformi 2 cc., Spiritus Rectificati 10 cc., Iodoformi 1 grm.—Iodi 50 centigrm., Etheris 15 cc., Acidi Carbolici 15 cc., Creasoti, 10 cc., Spiritus Rectificati 60 cc.

The number of quack remedies said to cure consumption is practically unlimited. It would be waste of time to discuss them.

Some drugs appear to exercise some influence for good upon the progress of the disease,—the **hypophosphites**, for example. We believe that we have seen some improvement resulting from the administration of **Calcis Hypophosphitis gr. iij, ter die.** The *rationale* of its action appears to be obscure. In view of the connection between the growth of the *Bacillus tuberculosis* and the production or spread of phthisis, any treatment to be really a treatment of the disease should aim, one would imagine, in producing such a condition of the lung-tissues or of the tissues generally as would prevent growth of the specific micro-organisms even after their introduction, or would prevent their introduction altogether. An alteration in the chemical composition of the lung-tissue might effect this; in what such an alteration is to consist and how it is to be brought about are questions for the future.

Syphilitic Disease.—It has been already pointed out that syphilis may enter into the causation of phthisis, by reducing the strength and power of resistance to the disease in those who have no hereditary tendency to it, as well as in those who have that tendency well marked. Syphilis may also arise in one who is already phthisical. In all of these cases, however, it is uncertain whether it in any way modifies or interferes with the development of the ordinary pulmonary lesions, although it has been stated, and this is our own experience, that the course of the disease is shorter under such circumstances. There is, however, little doubt but that the syphilitic virus may produce distinct lesions of its own in the lungs, quite independent of ordinary phthisis. This is looked upon as a rule as a late manifestation of the disease, but there are instances when in comparatively early syphilis one may suspect an implication of the pulmonary tissue. Three forms, at least, of pulmonary lesion due to this disease have been described. There is little difference between them histologically, namely: Firstly, gummata or syphilomata—these, when large and isolated, are uncommon and late lesions. They may be of the size of a chestnut or even larger, or they may not be larger than miliary tubercles (?). To the naked eye they often appear pale reddish in colour, ill defined about their edges, very tough on section, without marked signs of softening anywhere, not infrequently connected with fibroid thickening elsewhere in the lung, with dilated bronchial tubes, with emphysema in the neighbourhood, and with fibrous scars on the pleura. In structure they manifest, with more or less regularity, the ordinary structure of a gumma, viz. a central, degenerated, or caseous zone, surrounded by fibrous tissue, at the outside of which is the growing edge, fibronucleated, inflammatory tissue. As, however, a large gumma is generally made up by the coalescence of smaller ones, the fibroid tissue appears irregularly throughout the mass. The second and far more common form of lesion consists in a much more diffused fibroid induration of the lung-tissue, which may indeed be present in conjunction with distinct gummata.

This attacks the perivascular and peribronchial fibrous tissue as well as the alveolar walls, and produces thickening of the arterial walls, chiefly of the inner coat, which may go on to partial or complete obliteration of the vessels; thickening of the bronchial walls with moderate dilatation of the tubes; thickening of the alveolar walls, with gradual obliteration of the air-cells, and the production of greatly thickened fibrous septa in the portions of the lung attacked. The new inflammatory tissue is, indeed, similar to gummatous tissue, in that it consists primarily of a very vascular fibro-nuclear structure, which on the one hand tends to form distinct fibrous tissue, and on the other hand to degenerate and soften; the former tendency, however, is the most marked. In an account of certain cases of syphilitic disease of the lung, before the Pathological Society in 1877, Dr Goodhart gave an excellent summary of what appeared to him to be the distinguishing features of this fibroid induration, and with the majority of his points few can take exception. They may be thus summarised: That the lesion is prominently fibrous and not tubercular, that it generally attacks the base or the root of the lung and not the apex, that it is often associated with peculiar puckerings of the pleura, that it leads to gangrene and not to molecular or cheesy changes, that it is less evenly spread from lobe to lobe than is the case with chronic pneumonia and with the condition due to chronic pleurisy (p. 235), that it is nodular rather than diffuse, symmetrical and not unilateral, that the bronchial tubes are not so dilated as in anthracosis, and that the thickening is greater. This author seems to draw no distinction between the two preceding forms of the lesion.

Dr Green considered that the primary seat of the syphilitic affection is about the interlobular bloodvessels, and that this, coupled with its vascularity, is the chief diagnostic criterion microscopically of the specific lesion. The peribronchial and alveolar growths are subsequent to the perivascular changes.

A third form of pulmonary syphilitic lesion is described as occurring chiefly in congenital syphilis, in children up to ten or twelve years of age. It is called white pneumonia. It consists of a local or diffuse rapid proliferation of the interalveolar fibrous tissue resulting in thickening. The alveoli are filled with cells (catarrhal pneumonia), which after a time undergo caseous change, and are got rid of by absorption or expectoration. The result is chiefly manifested in great increase of the interalveolar tissue and of the alveolar walls, the alveoli themselves being partially or entirely obliterated.

The diagnosis of syphilitic disease such as we have described, according to my late colleague, Dr A. B. Shepherd (Gulstonian Lectures of 1876), depends upon the following points: Absence of any hereditary history of consumption; distinct history of syphilitic taint; cough, accompanied by very slight, if any, expectoration; dyspnœa, more intense on exertion than the dyspnœa of ordinary phthisis, and more resembling that of emphysema and chronic bronchitis; great emaciation; marked clubbing of the nails; absence of diarrhœa, or of any symptom of intestinal ulceration; absence too of any signs of granular kidney,—the patients being of well-made bodily structure, and have enjoyed good health for some time after the invasion and ordinary results of the syphilitic poison. The physical signs in such cases appear to be less than in ordinary phthisis. Slight dulness or even hyper-resonance, no marked bronchophony, slight superficial creaking, perhaps at both lower lobes behind, no distinct mucous râles.

Hæmoptysis, even fatal, has been observed; indeed, this symptom has been known to occur very early in syphilis, and we have seen a case, in an apparently healthy young man, in which it was of the most intractable character, and which came on about six weeks after the initial syphilitic lesion appeared.

The treatment to be attempted as soon as a diagnosis has been made is an anti-syphilitic one, with mercury and iodide of potassium, combined with tonics and cod-liver oil.

CHAPTER IX

DISEASES OF THE PERICARDIUM

I. Pericarditis [inflammation of the sero-fibrinous investment of the heart]

Conditions under which it arises:

1. *In rheumatism.*—The most common cause of pericarditis is undoubtedly acute rheumatism. It may appear as a complication, in any part of the course of this affection, and may even precede the swelling of the joints. In some cases it is the chief if not the sole manifestation of a rheumatic attack. The severity of the rheumatism, and the likelihood of the appearance of pericarditis, are not therefore in direct proportion. Statistics which show the relationship in point of frequency between the two affections have been often compiled. In a recent analysis of nearly 700 cases of acute rheumatism, mostly treated in his wards, Dr Church shows that 50 male and 26 female patients were attacked by pericarditis, or nearly 11 per cent.; as, however, the number of male and female patients included in the total were about equal, the percentage of males attacked was considerably the greater. This percentage is lower than Dr Sibson's, whose 326 cases gave over 19 per cent., as well as Dr Pye-Smith's, whose 400 cases gave 17·5 per cent. attacked. Dr Church believes that pericarditis is less common as a complication of acute rheumatism now than it was formerly.

2. *In acute Bright's disease.*—In acute tubal nephritis or acute Bright's disease there is a pre-

disposition to inflammation of serous membranes, including the pericardium. Pericarditis arises in the course of the affection as a complication, and so its relation to this disease is different to its relation to rheumatism. In rheumatism it must be considered as an initial manifestation of the disease. Scarlet fever, smallpox, and the other specific fevers, probably belong to the same class of predisposing causes of pericarditis as acute Bright's disease.

3. *In the specific fevers and pyæmia.*—In these affections, pericarditis arises as a late sequela, and occurs with the inflammatory affections of joints, the periosteum, and the bones.

4. *By extension of inflammation.*— When the neighbouring tissues and organs are the subjects of inflammation, the pericardium, too, may become similarly affected. This is the case sometimes in cervical cellulitis, pleurisy, and inflammation of the mediastina. The irritation of new growths, too, *e. g.* tubercle or carcinoma, within it or in the neighbourhood, may excite inflammation of the pericardium.

5. *Traumatic pericarditis.*—Perforating wounds, broken ribs, blows, rupture of neighbouring organs into the sac from pressure, *e. g.* of the stomach; or of aneurysms, cysts, or abscesses, may set up pericarditis.

6. *Idiopathic pericarditis.* — When pericarditis arises apparently independently of any of the above-mentioned causes it is called idiopathic, and under such circumstances *exposure to cold* is sometimes suggested as a cause, or if that be out of the question, and if there be a considerable amount of pneumonia and pleurisy prevalent, *epidemic influence* may be looked upon as a possible factor in its causation. Pericarditis as a constitutional and specific affection, apart from any other disease, has not yet been demonstrated.

Pathology.—The pathology of pericarditis is very similar to that of pleurisy and other serous inflammations, and the varieties of the affection are consequently similar to those which have been already described as forming the varieties of pleurisy.

1. **Dry pericarditis.**—The simplest form of pericarditis occurs in very minute patches upon the internal surface of the membrane over which the endothelium undergoes proliferation. Slight thickening results, and the so-called *tendinous* or *pearly spots* are produced; some, however, deny the inflammatory origin of these appearances. From this, the slightest form of the inflammation in which there is no evidence of increase in the pericardial fluid, to the two varieties of the affection to be next described, there are naturally many gradations.

2. **Pericarditis with copious serous effusion**, and 3, **pericarditis with sero-fibrinous exudation**.—These two forms of the inflammation cannot be considered apart as there is no line of demarcation between them. In some cases the fibrin, in other cases the serum, is in excess, but the one variety insensibly passes into the other. The chief difference between pericarditis and pleurisy as regards the effusion is that, in the former the effusion is more often fibrinous than serous, whilst in the latter the reverse is the case.

The stages in the inflammatory action in pericarditis are as follows:—First of all there is *injection of the blood-vessels* with *infiltration of the tissue* of the pericardium. The surface of the membrane becomes dim and loses its shining appearance. Small *elevations* of the membrane then occur from the growth of lymphoid nodules from the endothelium, and these form the beginning of what afterwards become adhesions. Small *extravasations of blood* take place here and there. If the inflammation goes no farther than this we have a dry pericarditis which may or may not end in small local adhesions from the continued growth of the villous processes already spoken of.

As a rule, however, *exudation* goes on *into the pericardial sac*. The solid fibrinous part of this is deposited upon the pericardium whilst the serous fluid is free within its sac, but is not, at first at all events, at the lowest level as one would expect; this is occupied by the heart itself and the fluid collects above. The principal exudation is found at the base of the heart at the root of the great vessels, and anterior, where the inflammation seems most often to begin. Later on, the serum finds its way to the lower part of the sac, and may even distend it to such an extent that the lungs and other parts in the neighbourhood may be compressed.

When the fibrin is in excess, which is the rule in the pericarditis of acute rheumatism, the heart pre-

Fig. 45.—A heart and the sac of the pericardium, the latter opened to expose the heart. Both internal surfaces of the pericardium are covered with a thick fibrinous deposit arising from sero-fibrinous pericarditis. Part of the deposit is regularly reticular, other parts are granular and warty. (From a photograph of a specimen in the museum of St Bartholomew's Hospital.)

sents a rough and shaggy appearance, quite characteristic.

This appearance has been likened to that presented by their inner surfaces on pulling apart two pieces of bread and butter which have been pressed together (Fig. 45). The deposit may be even more coarsely honeycombed.

Adhesions of the pericardium to the heart may result from these two varieties of the inflammation as well as from the first, absorption of the more fluid portion of the exudation taking place. The adhesions may be partial, or so complete as to render the dissection of the membrane from the heart post mortem a matter of much difficulty.

In comparatively rare cases the fibrinous deposit is deeply blood-stained from extravasations which have taken place in consequence of rupture of the dilated vessels of the pericardium itself, or of those which have been developed in the inflammatory false membrane; the condition produced is called *hæmorrhagic pericarditis*; it never arises unless the inflammation is very intense.

4. **Pericarditis with purulent effusion.**—Pus in the pericardium forms the fourth main variety of pericarditis. It is a rare result of sero-fibrinous inflammation of the membrane. It may appear also in pyæmia. What exactly determines its formation in the former case is impossible to say. The appearance of pus in cases of pericarditis is much less common than is empyema in cases of pleurisy. In extremely rare cases the purulent effusion may become putrid, gas may be evolved, and a condition of pneumo-pericardium may be developed.

Effect of pericarditis upon the heart.—The heart may be seriously affected in pericarditis. Unless the exudation is small, some interference with the cardiac action must occur.

    a. In cases of excessive exudation (p. 297).—Excessive exudation tends to produce softening of the

muscular tissue, and dilatation of the heart; or if dilatation does not occur, then atrophy.
b. *In the case of adhesions of the pericardium to the heart.*—Adhesions also interfere with the proper cardiac movements, hence arises, in most favorable cases, hypertrophy of the cardiac walls. This, however, is not without some dilatation, and the further history of the heart shows that the hypertrophy may ultimately give place to fatty degeneration and atrophy as in the former case.
c. *In case of excess of the deposit upon the heart.*— Excessive deposit upon the heart occurs in sero-fibrinous exudation, forming a false membrane covering the heart. This may not be absorbed and got rid of, but may remain; may be converted into a hard fibrous tissue which tends to hamper the muscular action.

Myocarditis may arise in connection with the inflammation of the pericardium, if it be of a very severe type.

Symptoms and course.—The symptoms of pericarditis, when it arises, as is mostly the case during the course of acute rheumatism, are generally obscured by the symptoms of the joint affection. But if **pain in the præcordial region** arises or suddenly increases, inflammation of the pericardium may be looked for, especially if the pain is accompanied by **palpitation of the heart** and dyspnœa. Sometimes there is an **increase in the febrile symptoms**, shown by a marked rise in the temperature and a rapid pulse. Many or even all of these symptoms may, however, be absent. A systematic examination of the whole præcordial region with the stethoscope at least once a day during the course of rheumatism is therefore necessary.

If the exudation is excessive the **dyspnœa** is very intense, the patient exhibits great restlessness and is unable to lie down, but often sits up and leans forward in bed. The pained and terrified look upon his countenance is, one might almost think, characteristic. If lying down, his decubitus is on the left side, as in that way the pressure upon the right lung is diminished.

Dysphagia.—Some difficulty in swallowing is experienced from pressure of the distended pericardium upon the œsophagus.

Delirium.—There is sometimes marked delirium; so marked, indeed, is it that pericarditis may be mistaken for the delirium of tubercular meningitis in young people. We have seen the mistake made.

Cyanosis.—Blueness of the face, lips, and extremities in gradually increasing severity may occur from obstruction to the pulmonary and system circulation.

The pulse sometimes presents the characters of the **pulsus paradoxus** (Fig. 46).

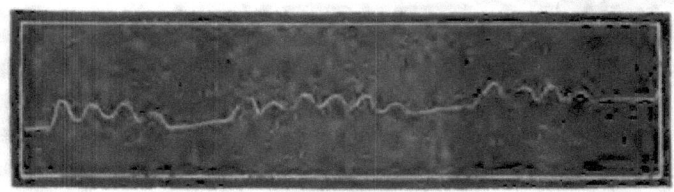

Fig. 46.—Diagram of pulsus paradoxus taken in a case of persistent pericarditis. (S. West.)

It should be remembered that pericarditis of a more or less extensive character arising in the course of renal disease, tubercle, and the like, may often be quite unaccompanied by symptoms.

Terminations:
i. **In recovery.** Pericarditis usually terminates in apparent recovery, especially when it occurs in the course of acute affections, such as rheumatism.

ii. **In incomplete recovery.** The acute affection may become chronic. This may happen in rheumatic fever. It is more often subacute or chronic from the beginning when the affection arises in the course of other diseases. Complete recovery from chronic pericarditis is rare. Relapses are common. The most common sequelæ of the chronic affection are: (*a*) Adherent pericardium; (*b*) dilatation of the heart; (*c*) hypertrophy of the heart; (*d*) atrophy and fatty degeneration.

iii. **In death.** (*a*) *In the acute stage,* (1) from gradual stoppage of the heart from pressure of the effusion. (2) From the complication of pericarditis with pneumonia and pleurisy. (3) From the severity of the initial disease, *e. g.* in pyæmia.

(*b*) *In the chronic stage,* (1) from gradual œdema of lungs, or (2) from complications and sequelæ, *e. g.* some of those above mentioned.

Physical Signs:

On inspection, (*a*) there is præcordial bulging, which, however, is chiefly marked in children.

(*b*) An alteration in the position of the apex-beat, as well as in its extent and force. The cervical veins may be distended.

(*c*) ? Bulging of the left lung above the clavicle.

On palpation, (*a*) the apex-beat is felt to be weaker, although more diffused than natural, and may be hardly felt at all unless the patient leans forward.

(*b*) Friction fremitus over the præcordia is sometimes distinctly felt.

On percussion.—The most characteristic feature in the percussion dulness is *its increase* first of all *upwards*, the upper limit being a line on a level with the first rib or first interspace, extending from a little to the right of the sternum to an inch or more to the left of that bone. After this, as the fluid increases the pericardial sac becomes distended around the heart, which is said by some observers first of all to *sink* in the fluid, and in consequence, on a level with the lower attachment to the diaphragm, the *transverse area of dulness is greatly increased* below, so that it may extend on the right to the nipple and on the left to the anterior axillary line.

The dulness after a time, in cases of extreme effusion, will be found to extend from the line above mentioned, on a level with the first rib or interspace above, to the transverse nipple line below. This area has been compared to an obtuse-angled triangle, having its base downwards.

The dulness extends beyond the apex-beat.

The liver dulness is depressed.

Sometimes the cardiac dulness is unaltered or but little affected. This occurs under three conditions, either when the effusion is confined to the posterior aspects of the heart, or when the pericardium is more or less covered by emphysematous lung, or when the percussion-note of the trachea or bronchi is conducted forwards by the distended sac.

The three diagnostic criteria obtained by percussion should indicate with considerable certainty pericardial effusion.

i. The *acute* extension of cardiac dulness upwards and laterally at the base.
ii. The extension of dulness to the left, beyond the apex-beat, in the transverse nipple line.
iii. The alteration of the shape and extent of dulness on altering the position of the patient, *e. g.* when he stands the broadest diameter is at the diaphragm.

On auscultation.—At first there is (*a*) friction or rubbing to-and-fro sound, heard most often at the base of the heart, or at the edge of the cardiac dulness to the left. This sound may continue in spite of the presence of a considerable amount of fluid in the pericardial sac. When this becomes very great the sound is no longer heard. It may disappear also from absorption of the effused lymph, or if adhesions take place.

(*b*) In addition to the friction sound it is noticed that the cardiac sounds are feeble. The differences in the character, position, time, and extent observed between pericardial and endocardial sounds have been already indicated (p. 111).

Prognosis.—As to partial recovery the prognosis is distinctly favorable, patients seldom dying of acute pericarditis itself. When it arises as a late complication of affections, such as pyæmia, pericarditis cannot be considered as doing anything but hastening death. The sequelæ of the disease are very serious, adherent pericardium especially so, and the prognosis as to length of life of a patient must be guarded. It should be recollected that in many cases, especially of rheumatic origin, complete recovery may be looked for.

Treatment.—When pericarditis occurs in the course of rheumatism, the practice of applying a *blister* over the seat of pain or over the position of a pericardial rub or fremitus is a good one; instead of this, a *mustard poultice* or *leeches* may be applied, or *Ung. Hydrarg.* may be rubbed in; the latter practice, we believe, in spite of anything which has been aserted to the contrary, sometimes does good.

Opinions differ as to whether the rheumatic treatment should in all cases be pushed, but, at any rate, if the alkali plan is being carried out it should be persisted in and followed up energetically.

As regards drugs, *aconite* (tincture ♏ij to ♏v) is regarded as urgently indicated by some physicians. If there be any tendency to a weak action of the heart, *digitalis* (tincture ♏v to ♏x) may be given with advantage.

Iodide of potassium internally and *iodine* externally are recommended as helping to procure absorption of pericardial effusion when subacute. Even if these be useful, we put no faith in drastic purges or diuretics.

Tapping the pericardium, or paracentesis pericardii, should only be resorted to when it is evident that the patient will die if the pressure upon the heart and lungs is not speedily relieved. It is a serious operation, but need not be dangerous. In an account of 72 cases* collected by Dr Samuel West, in only one was the operation itself fatal. When paracentesis is required it should be done in the fifth left intercostal space, one inch from the sternum, with a fine trocar and cannula. The pneumatic aspirator may or may not be used. In purulent pericarditis the treatment which holds out the greatest chance of recovery seems to be free incision into and draining of the pericardial sac. Irrigation of the sac with antiseptic solutions may be combined with this treatment.

II. Adherent Pericardium

[A condition in which the pericardium, as a result of inflammation, becomes more or less adherent to the heart.]

Pathology.—The extent of the adhesion of the heart to the pericardium varies greatly—as has been pointed out in the last article—from a few bands to complete and firm investment of the whole organ. The thickness of the agglutinating material, its consistency and state of organisation, also vary. The most complete adhesions of the heart and pericardium are met with in young people. The effect upon the heart has been already indicated.

Symptoms and physical signs.—It is impossible to indicate any symptom which is at all characteristic of adherent pericardium. A rapid action of the heart, with a feeble and irregular pulse, may be

* 'Medico-Chirurgical Trans.,' vol. lxvi, p. 256.

present, with **shortness of breath** and a **tendency to blueness of the lips** and possibly dropsy, but these symptoms can hardly be said to be due immediately to the condition of which we are treating.

As with symptoms, so with physical signs, although several are mentioned as possible, yet not one is certain—indeed, if the pericardium is adherent to the heart only, we know of no physical sign whatever which would indicate the condition.

If, however, the pericardium is **adherent not only to the heart, but also to the chest wall in front** or **behind or to both**, there may be these signs:

i. A depression of the intercostal spaces at the position of the apex-beat, during the cardiac systole, and its cessation during diastole. This may happen,* so it is said, even although there be no adhesion of the pericardium to the chest wall anteriorly, but how, it is difficult to imagine. If there be no pleuro-pericardial adhesion the sign is absent.

ii. If there are pleuro-pericardial adhesions, the area of cardiac dulness and the position of the apex-beat are unaffected by a deep inspiration.

iii. There is systolic recession of the epigastrium.

iv. The jugular veins are distended during systole and collapsed during diastole.

III. Hydropericardium [hydrops pericardii]

[An abnormal collection of fluid within the pericardium of non-inflammatory origin.]

Causes.—It must be recollected that an apparent increase in the amount of fluid noted in the pericardial sac after death does not of necessity indicate anything but a serous transudation a few hours, it may be, before death has occurred. In other cases—

i. A collection of serous fluid in the pericardium is nearly always a part of a general dropsy, e.g. that arising in chronic kidney or lung disease.

* See Niemeyer, vol. i, p. 390.

ii. A local pericardial dropsy occurs when from any cause the veins of the heart itself are obstructed. This condition is observed in chronic lung disease, especially in bronchitis and emphysema, in the later stages of phthisis, and of valvular diseases of the heart.

iii. An increase of fluid takes place into the pericardium when atrophy of the heart occurs after hypertrophy, and also when the pericardium is adherent to the lungs and they contract, and by so doing distend the pericardial sac. Indeed, an increase in the pericardial fluid may occur in any case in which the intrapericardial pressure is permanently and gradually diminished.

Post-mortem appearances.—The amount of serous fluid within the pericardium is seldom anything but moderate, two to four ounces. It may, however, be much greater, and in such cases the heart is small, flabby, and pale.

Symptoms and physical signs.—With the exception that the presence of fluid within the pericardium tends to increase the embarrassment of the heart, already weak, and to increase the dyspnœa, already marked, dropsy of the pericardium cannot be said of itself to produce **any symptoms.**

As regards physical signs, they are similar to those of pericardial effusion, but the dulness is more often masked by the lung, adherent to the chest wall, being in front of the distended sac; præcordial bulging is seldom present. The effusion is of course not preceded by a friction sound.

Treatment.—The treatment must be directed towards the original affection, of which the dropsy into the pericardium is a symptom. Hence, one would expect some help from drugs which act upon the circulation generally; thus *tonics* such as iron and strychnine, *cardiac stimulants* such as digitalis, might be tried. *Diuretics* and purgatives may also be administered if the patient is able to bear them.

IV. Pneumopericardium [a condition in which air or gas exists in the pericardium]

Causes.—The causes of this somewhat rare disease are as follows:
1. Penetrating wounds, under which may be included paracentesis pericardii.
2. Perforation of the pericardium in diseases of other organs of the neighbourhood containing gas, *e.g.* of the stomach, carcinoma or ulcer of the œsophagus or of the lung.
3. In purulent pericarditis, gas may, in very rare cases, be generated.

Post-mortem appearances.—It is almost impossible that air should exist within the pericardial sac, without there being at the same time some fluid, fibrinous or purulent, also present. The mere presence of gas or débris from the encroaching growth would of itself be quite sufficient to set up inflammation, and so cause effusion.

Symptoms and physical signs.—When perforation of the pericardium takes place, a condition of collapse, not at all characteristic however, occurs, from which, if the patient recovers, physical examination of the chest reveals signs of a highly characteristic kind.

On inspection—(*a*) There is bulging of the præcordia.

(*b*) There is diminution of force or absence of the apex-beat.

On percussion.—There is hyper-resonance over the præcordia, sometimes of distinctly metallic character, when the patient is lying down upon his back; when he sits up and leans forward this disappears.

On auscultation.—*a.* A clear, splashing sound caused by the movements of the heart in the fluid is heard when air and fluid are present within the pericardium; this sound is audible at considerable distances from the patient; in a case mentioned by Niemeyer it was audible to the room-mates of the patient at the other end of the ward.

b. **Succussion sound** may be audible under similar circumstances.

c. The heart-sounds are very faint, but when audible may have an amphoric character.

d. Friction sounds may be heard which may or may not have a metallic echo.

Prognosis.—Pneumopericardium, unless traumatic, is almost invariably and quickly fatal. The collapse is very great, and even if that be recovered from, the after-inflammation is intense. Traumatic pericarditis is not, however, by any means hopeless.

V. New Growths of the Pericardium

α. **Tubercle** in the pericardium is, as a rule, part of a general tuberculosis, and as such is not likely to produce symptoms which can, over and above the symptoms of tubercle elsewhere, indicate the implication of the pericardium in the general infection.

Tubercles may also form locally, so it is said, as a result of chronic pericarditis. They do not, however, give rise to any fresh symptoms.

β. **Carcinoma.**—Carcinoma generally arises within the pericardium as a secondary affection (i) by extension from the glands or bones in the neighbourhood. The amount varies from small nodules more or less flattened, to an almost complete and thick capsule, as we have seen on one occasion. (ii) As a part of a general cancerous infection after the extirpation of an external cancer.

This disease is not likely to call for special diagnosis or treatment.

CHAPTER X

DISEASES OF THE HEART

I. Hypertrophy of the Heart [or an increase in the muscular tissue of the heart]

Etiology.—Hypertrophy of the heart is almost invariably found to be a conservative and beneficial change. It occurs whenever some extra stress is put upon the heart to overcome some increase of the peripheral resistance, or some obstruction to the flow of blood nearer the central organ itself. The causes which produce it may therefore be summarised as follows:

1. **Obstructive valvular diseases,** whereby the flow through any of the orifices is obstructed or delayed by constriction. Hypertrophy in this case may at first, at any rate, be confined to one of the chambers; this is the case, for example, in mitral stenosis.

2. **Incompetence of the valves of the heart** also ultimately produces the same effect, and the hypertrophy in this case may also be at first confined to one chamber. It is probable that in both kinds of valvular disease, however, dilatation either precedes or accompanies the increase in the muscular tissue.

3. **Obstruction to the circulation in the large arteries** produced by (*a*) a disease of the coats of the main arteries, *e. g.* atheroma, or by (*b*) aneurysmal dilatation; (*c*) constriction beyond the valves; or by (*d*) pressure diminishing its calibre.

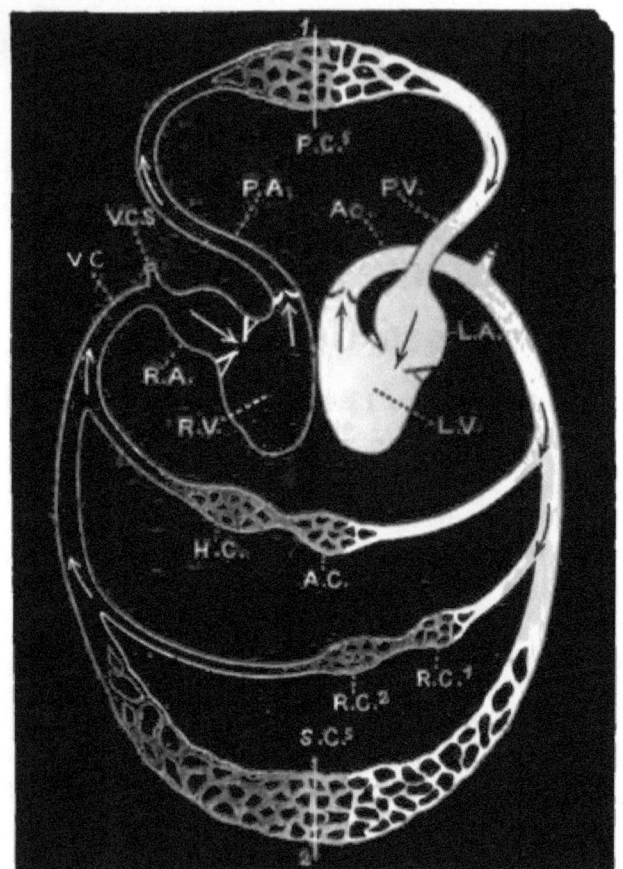

Fig. 46.—Diagram of the circulation to illustrate the mechanism of the dilatation and hypertrophy of the heart. R.A., right auricle; R.V., right ventricle; P.A., pulmonary artery; P.C.*, pulmonary capillaries; P.V., pulmonary veins; L.A., left auricle; L.V., left ventricle; Ao, aorta; C.C., cerebral circulation; A.C., circulation of alimentary canal; H.C., hepatic; R.C.¹ and R.C.², renal circulations; S.C.*, systemic capillaries; V.C. and V.C.S., inferior and superior venæ cavæ.

4. **Increased peripheral resistance**, due to the contraction of the arterioles, or to obliteration of capillaries, the former being exemplified in the hypertrophy of the left ventricle in chronic Bright's

disease, and the latter in hypertrophy of the right side of the heart in destructive disease of the lung, or in which the circulation through the lung capillaries is chronically interfered with. A glance at Fig. 46 will explain this.

5. As a necessary consequence **of dilatation of the** heart. We have already mentioned that this is the case in valvular diseases; so also must it be, if the circulation is to go on, in dilatation of the heart arising from any other cause unless very temporary.

6. As a consequence of increased **muscular exertion.** This is the case with athletes, and with those who ply some very laborious occupation. It must be remembered, however, that not infrequently there is some little valvular defect to account for hypertrophy in not a few of such cases. Only repeated and careful auscultation can eliminate that possibility.

Under this same head may be considered the occurrence of hypertrophy in cases of long-continued palpitation. In the examination of a large number of dyspeptic females in the casualty department of St Bartholomew's Hospital some years ago we found that the relationship between palpitation of the heart and dilatation was very constant. In some cases a very considerable amount of dilatation was not uncommon.

7. **Plethora** is often regarded as a direct cause of hypertrophy. With reference to this reputed cause, however, it must be borne in mind that a condition of constant plethora is almost invariably associated with organic disease of the kidney or liver. The usual sequence of events is probably this: indigestion, palpitation, dilatation, hypertrophy.

Morbid anatomy and pathology.—On examination of hypertrophied hearts after death it is at once obvious that the hypertrophy is either **partial or general.** At one time all four chambers may exhibit hypertrophy of their walls, at another time three only, and sometimes the hypertrophy is entirely one-sided, and again, as before mentioned, but in rare cases one chamber alone may show a distinct increase in the thickness of its muscular walls.

With the hypertrophy dilatation is almost always present; this is known as **excentric** hypertrophy. Two other terms are used to express conditions which

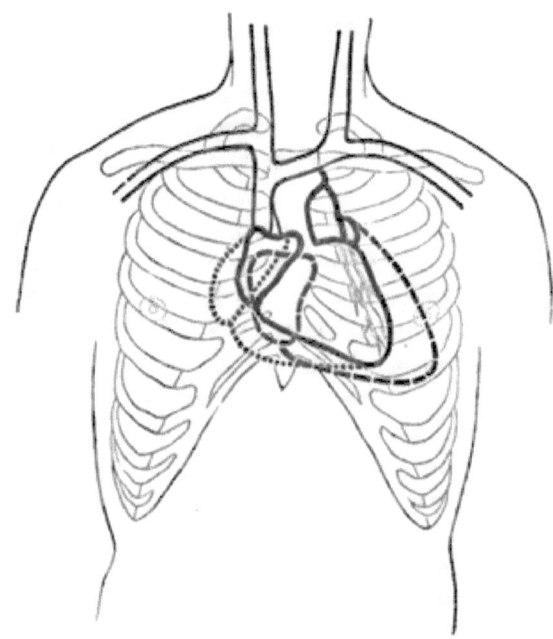

Fig. 47.—Diagram of the enlargement of the heart to the right and to the left in excentric hypertrophy of the right or left sides respectively. The uninterrupted line indicates the normal heart; the small interrupted line indicates right hypertrophy and dilatation; the line less interrupted indicates left hypertrophy and dilatation.

if they ever exist during life are at any rate very rare; these are **simple** hypertrophy when the walls are thickened without any increase in the capacity of the chambers, and the other **concentric** hypertrophy, that is to say, hypertrophy at the expense of the chambers or hypertrophy with contraction.

Changes produced by hypertrophy.

Changes produced by hypertrophy.—In hypertrophy, the heart is changed in **weight**, in **size**, in **the thickness of its walls**, in shape, and **in position**.

Its **weight** is increased from nine or ten ounces to twenty or thirty or more; the **size is increased**, may be doubled or trebled.

Thickness of walls is increased thus:

	Normal.	Hypertrophied.
Left ventricle	4 to 4·5 lines.	5 to 12 or 18 lines.
Right ,,	1·75 to 2 lines.	2·5 to 9 lines.
Right auricle	1 line.	2 lines.
Left ,,	1 line.	3 lines.

Position.—When affected with total hypertrophy the heart has a tendency to incline downwards, and with the apex to the left and base to the right; if the left is affected alone the prominence is to the left, and if the right is affected alone the increase in area is to the right side (see Fig. 47).

Shape.—If the hypertrophy is complete, the shape of the heart is that of an obtuse-angled triangle; if the left side is hypertrophied alone, the heart is elongated and conical; if the right side alone, it is broad and spherical.

Nature of the hypertrophy.—Probably numerical, that is to say, that the number of the fibres is increased.

Symptoms and physical signs.—It is practically impossible to separate the symptoms of hypertrophy of the heart from the diseases which give rise to it. Indeed, hypertrophy by itself produces few symptoms, but there may be—

- Palpitation, more often, however, objective than subjective, and seldom complained of by the patient, even when the amount of the hypertrophy is considerable.
- Shortness of breath, due possibly to encroachment on the space which should be occupied by the lungs.
- Some **tightness or fulness in the chest.**

The **pulse** is full, strong, and rapid, and is felt plainly in the small arteries.

A **visible pulsation in the vessels of neck**.

The **face may be red** and the eyes glistening.

With or without these symptoms the patient enjoys good health, but is liable to fluxions to the brain or lungs and to cerebral hæmorrhage.

On physical examination of the chest.—On inspection, (*a*) there is bulging of the præcordia. (*b*) The apex-beat is extensive and forcible, extended to the left and downwards, especially if the left side alone be affected; and marked in the epigastrium, and to the right if the right side is affected, but also to the left, but not so much downwards.

On palpation, the impulse is felt to be strong and jarring.

On percussion, the cardiac dulness is increased downwards and to the left and also to the right, so that, when the whole heart is hypertrophied, the dulness is increased transversely and longitudinally, but not, as a rule, upwards, the area retaining its oval shape.

On auscultation.—When the stethoscope is applied to an hypertrophied heart, the head of the auscultator is observed to be distinctly raised at each cardiac impulse. The heart-sounds are increased in loudness, and are heard distinctly over a more extensive area than natural. Nearly always, murmurs indicative of valvular disease are present. A whiz is heard on auscultation of the arteries when they are slightly pressed upon by the stethoscope.

Diagnosis.—Excentric hypertrophy of the heart may be mistaken for:—Pericardial effusion, mediastinal tumours, aneurysms of the aorta, localised pleural effusion, consolidation of the lung anteriorly; but chiefly for the first-mentioned condition. In the diagnosis reliance must be placed upon a careful consideration of the physical signs (for pericardial effusion, see p. 301).

If there be a small pericardial effusion with acute dilatation of the heart the diagnosis becomes a matter of great difficulty.

Treatment.—As hypertrophy of the heart is a conservative effort of nature it obviously does not require treatment.

The conditions which have given rise to it must be sought for and, if necessary, treated.

Some general rules, however, may be laid down:— Excitement, great muscular exertions, over-eating and drinking, and all such conditions which tend to produce distension of the stomach with gas must be avoided.

Purgatives are often necessary to keep the bowels freely open.

Applications of belladonna plasters are useful if palpitation is troublesome.

II. Dilatation of the Heart

[A condition of enlargement of the chambers of the heart, without a corresponding increase in its muscular substance.]

Etiology.—Dilatation of the heart is a problem in the physics of the circulation. It arises whenever for any length of time the normal relationship between the propelling power of its muscle, the normal amount of the blood propelled at each contraction, and the resistance which the propelling power has to overcome in the arteries is disarranged. Inasmuch, however, as the main circulation is, as it were, a double one, viz. systemic and pulmonary (see Fig. 46), each with its own propelling half of the heart, it is manifest that the equilibrium of the forces which maintain the circulation of each part may be disturbed either in one part or the other. Of course, as the two parts of the circulation are continuous, one is not absolutely independent of the other part, but there is no doubt that if the obstruction occurs in the systemic circulation the left heart is at first and chiefly affected, and if in the pulmonary circulation the right heart is primarily and chiefly affected. Thus we may have dilatation confined to or chiefly present on one side of the heart. For several reasons the right heart is more frequently affected than the left. Firstly, because it is weaker than the left, and secondly, because sufficient obstruction to affect the propelling power is more likely to occur in the pulmonary than in the systemic capillaries.

It will be necessary to consider some of the chief ways in which the natural equilibrium of the factors which maintain the circulation is disarranged. A reference to Fig. 46 will greatly aid the explanation.

i. Firstly, by *an increase in the peripheral resistance.* If this be permanent, there is a gradual dilatation of the propelling chamber, but with the dilatation there is, as a rule, a corresponding hypertrophy of its walls, and the relationship between the factors, even if for a time disarranged, is again righted, as the heart, contracting with greater force, overcomes or rather counterbalances the obstruction. This, we may suppose, occurs in renal disease to the left heart and in emphysema to the right heart. The same sequence takes place if the *obstruction be nearer to the propelling force*, as, for example, in cases of *obstruction at the valvular orifices*. In both these cases, the first deviation from the normal may be the inability of the dilating chamber to expel its contents at each stroke. It appears contrary to reason to believe that, if this should increase little by little, a mere increase in the rapidity and force of the contraction could of itself, without a corresponding increase of cardiac muscle, produce a return to the normal relationship. This, however, is asserted by some. What really happens is either that the hypertrophy accompanies the dilatation, or that dilatation, unless very slight, produces all of its effects, until such time as the hypertrophy catches it up. Again, similar events happen if dilatation is due to *valvular incompetence*, and probably also in the cases in which the *total amount of the blood in the body is increased*.

ii. The balance between the constituents of the equilibrium, however, may be disarranged in a second way. *The resistance may remain the same, but the propelling power may be diminished;* this happens when from any cause the power of contraction of the muscular substance is diminished, either (a) constitutionally, *e. g.* in anæmia, in convalescence from acute diseases, or (b) locally, when the muscular substance is actually degenerated from fatty changes, inflammation, or the like.

iii. Again, the resistance *may be greater than normal, but the power may diminish, having been increased to or above normal.* This is the case when hypertrophy is succeeded by fatty degeneration of the heart fibres.

The first of the above conditions has been considered in the last section. The others are those to which attention is now being directed.

The question naturally presents itself, how long may the circulation be maintained and dilatation continue or increase? The answer as naturally divides itself into two sections. Firstly, the dilatation must be in nearly all cases very gradual, and must often be interrupted, possibly by a temporary energy in the muscular contractions, due either to a general improvement in health or to an extra stimulus supplied by medicines or what not, or the resistance may be temporarily diminished.

In all cases, however, the tendency must be for the dilatation to increase, however slowly, except in those in which the **want of tone in** the heart muscle is a temporary condition, and one certain to be recovered from.

This tendency is aided as time goes on by the effects of the increasing dilatation. Then comes a gradual weakening of the dilated chamber, from stretching, a diminishing amount of blood sent into the arteries, a rapid, irregular, and more and more inefficient contraction of the heart; increased venous pressure with serous effusions, diminished arterial pressure and small pulse.

At last a time comes when the chambers become so stretched that their walls are no longer able to contract upon their contents, and a condition of asystolism of Beau results, and the heart stops beating.

Although, as a rule, these steps are slow, yet what may be called accidental circumstances may hurry them, and an intercurrent bronchitis, for example, to which there is a tendency from the congested condition of the bronchial mucous membrane, may greatly aggravate the condition of a patient suffering from heart disease, and in whom, it may be, the hypertrophy is giving way to degeneration.

Post-mortem Appearances.—The dilated heart after death appears, as a rule, to be larger than natural, and frequently the cavities are distended with clot. The right ventricle is the chamber which most often exhibits dilatation, and the muscular substance is stretched and thinned. The tissue of the organ does not present under the microscope any apparent alteration from the normal, unless the dilatation be the result of fatty changes. Although frequently associated with valvular disease, yet the valves may be found competent to close the orifices in dilated hearts.

Symptoms.—Even whilst restricting our account

to uncompensated dilatation of the heart it is not easy to say what symptoms are due directly to dilatation and what to its cause. The chief, however, may be stated as follows:

Palpitation.—Palpitation is more often troublesome than it is when the hypertrophy continues to be compensative, although the impulse of the heart is much less forcible and diffused. As before mentioned, the symptom is seldom much complained of as long as the hypertrophy of the heart is able to maintain the circulation in a fairly normal condition. It is due first of all to the endeavour of the heart to make up in the frequency of its contractions for the diminution in the amount of blood which it discharges at each contraction. The symptom is felt too as long as dilatation exists without hypertrophy. It is therefore frequently present when the muscular substance is weakened from any cause whatever.

It consists of a disagreeable fluttering and throbbing in the præcordial region, increased on exertion and after meals.

As the tissue of the heart becomes still weaker, irregularity is added to rapidity, and the impulse gets less and less forcible and often imperceptible, but the subjective sensation may continue or may be felt even more plainly than before.

Dyspnœa.—Considerable difficulty in breathing, increased on the slightest exertion, is experienced at first when walking uphill or upstairs only, but gradually progressing until the subject of the disease cannot rest unless he sits up in bed or on the side of the bed. He sleeps little at night.

This symptom is to be attributed to the gradual congestion of the pulmonary vessels from the obstruction to the circulation from in front, and from imperfect contraction of the right ventricle.

Cyanosis.—The patient after a varying time exhibits a certain blueness of the lips and tongue, and the venules of the face are dilated. This blueness is due first of all, to imperfect oxygenation of the blood from delay in the circulation. Afterwards, there is added to this, a permanent dilatation of the veins.

Dropsy.—Effusion of serum into the tissues is a natural and invariable accompaniment of the venous congestion. At first it is generally confined to the lower extremities, but later on it is more or less marked elsewhere, and extends to the subcutaneous tissue of the back, to that of the scrotum and face; later still there are dropsical effusions into the serous sacs, into the peritoneum, pleura and pericardium.

Congestion of internal organs, *e. g.*

Of the liver, with enlargement, more or less painful, and sometimes pulsation, slight jaundice, hepatic breath.

Of the kidneys, with scanty, high-coloured urine of high specific gravity, depositing abundant urates on standing, and containing some albumen, varying from a mere trace to $\frac{1}{4}$ or $\frac{1}{3}$.

Of the brain, with drowsiness, apathy, wandering at night, bad dreams.

The Pulse is small, rapid, and irregular, often intermittent.

Physical signs are not marked. The apex-beat is diffused, probably displaced, perceptible in the epigastrium, or to the left, but is feeble and often neither to be seen nor felt. There is an extension of the dulness to the right or left, or in both directions, much in the same way as in hypertrophy. The heart-sounds are weak; the first especially so.

It should be remembered that there are sometimes added the physical signs of incompetence of the tricuspid valve, shown especially in pulsation of the jugular veins, and pulsation of the liver, with or without a systolic murmur, loudest over the ensiform cartilage.

Course and terminations.—1. If the dilatation is followed or compensated for by an increase in the muscular tissue of the heart, the symptoms which have arisen in consequence of the dilatation disappear, and, for a time at any rate, the patient is free from the shortness of breath and palpitation or has them to a very slight degree.

2. When the dilatation arises from a permanently degenerated condition of the muscle, even although it have been hypertrophied for a time, then the prognosis as to anything but a temporary relief of symptoms is unfavorable, and, indeed, the tendency is distinctly from bad to worse. The obstruction of the circulation in the lungs increases, and the lungs themselves gradually become waterlogged. It is astonishing, however, how long patients live on in this condition, varying from day to day. Œdema of the lung, or some intercurrent affection, usually after a time terminates the case.

3. When the dilatation is due to a weakened condition of the heart muscle, which is not of necessity permanent, then it is certain that the patient may be very materially benefited by careful treatment, and recovery may take place. Such conditions are noted in anæmia, prolonged dyspepsia, and in convalescence from fevers and other debilitating illnesses.

One or two considerations with respect to such cases may be added, viz.:

i. When dilatation to a moderate amount arises from want of muscular tone, the arterial tension is probably diminished, and so the weakened ventricle may be sufficient to maintain the circulation.

ii. Dilatation of the heart, although sudden death may arise from it, is not often a hopeless condition unless the muscular tissue is permanently weakened by actual degeneration.

Treatment.—Naturally falls under the three heads representing the three conditions under which dilatation arises.

1. When the hypertrophy of the heart follows dilatation, so that it may be compensated for, little treatment is required except—

   Entire **rest**.
   Abstention from excitement of all kinds.
   Good nutritious diet, frequent light meals.
   Tonics, such as iron and strychnine.

2. When dilatation arises from debility of the heart muscle as a part of a general debility,

   Change of air and scene, sea air often being especially beneficial.
   Gentle exercise, to be gradually increased as the strength increases.
   Good nutritious diet—milk, eggs, alcohol in the form of light wines—Burgundy, Rhine wine and the like.
   Tonics, of which especially iron in the form of tartarated iron (gr. v), citrate of iron and quinine (gr. viij) or citrate of iron and ammonia (gr. viij). Easton's syrup (ʒj), quinine (gr. ij), Parrish's chemical food, and cod-liver oil (ʒij) and strychnine.

3. When dilatation arises from degeneration of the muscular substance of the heart, treatment is of great importance.

   Absolute rest in bed in a suitably warm and not draughty room, and freedom from excitement of all kinds are necessary.
   Good nutritious diet, which should be liquid, chiefly milk, arrowroot, beef-tea, or soups; a certain quantity of good wine or spirit is often necessary.
   As to **drugs**, (1) gentle purgatives—senna, rhubarb, or cream of tartar, to counteract constipation.

2. Cardiac tonics, especially digitalis (Tr. ♏vij to ♏x), strychnine, and iron.

Symptoms must be treated as they arise.

If the congestion of the lungs be great and œdema imminent, bleeding from the arm should certainly be resorted to. If the urine be scanty, diuretics, *e. g.* in the form of Pil. Digitalis co., which contains Digitalis leaves, Pulv. Scill. and Pil. Hydrarg., is greatly to be recommended.

For **dropsy**, purgatives, such as jalap, scammony, and cream of tartar if they can be borne.

Local applications, in the form of hot poultices, over the chest or abdomen; stimulating liniments and the like are often grateful to the patient.

III. Atrophy of the Heart

[An abnormal smallness or diminution in the size of the heart.]

Etiology.—i. In the first place the heart may be congenitally small, so as to be in the adult no larger than that of a child. It is combined with retarded development of the body in general, and of the sexual organs in particular. It occurs chiefly in females.

ii. The heart is atrophied in cases of general wasting, as, for example, in phthisis, in carcinoma, and in old age. It also may occur in any condition in which there is insufficient nourishment supplied to it, as in cases of interference with its blood-supply through the coronary arteries.

iii. Prolonged pressure, as in pericardial effusion or chronic pericarditis, sometimes causes the heart to become atrophied.

Post-mortem appearances.—The congenitally small heart is as above mentioned like that of a child; it is a normal heart in miniature. When it is the result of disease it is also small both in its cavities and in the thickness of their walls. It is generally anæmic and flabby, but may be hard and fibrous. The fat disappears from the sulci. Within the pericardium is an abnormal amount of fluid. The coronary vessels are often dilated and full of blood.

Symptoms.—i. *Congenital smallness* of the heart is said by good authorities to give rise to—

Frequent fainting attacks.
General malnutrition and hindered development.
Palpitation.
Chlorosis and anæmia.

ii. *When acquired*, (*a*) as part of a general marasmus—

The pulse is small; the apex-beat is feeble.
The sounds of the heart are feeble.
There is venous engorgement more or less marked, with or without—
Cyanosis and dropsy.

(*b*) When due to *local disturbance to the contraction* of the heart there are, in addition—

Palpitation and
Dyspnœa, and the other signs are exaggerated.

(*c*) When the result of *fatty degeneration*—

Cyanosis and dropsy, with feeble pulse, cold and blue extremities, diminished cardiac dulness, and weak heart-sounds are present.

Treatment.—Muscular exercise of any violent kind is to be avoided. Good food, and good general hygienic conditions; wine in moderation, and tonics, both general and local,—quinine, iron, strychnine, and cod-liver oil—may be employed.

IV. Endocarditis [inflammation of the serous membrane lining the heart]

Etiology.—Inflammation of the endocardium arises under many of the conditions which have been described as giving origin to pericarditis.

(1) **Rheumatism.**—As of pericarditis, rheumatism is by far the commonest cause of endocarditis, and, indeed, this affection is even more likely to arise in the course of acute rheumatism than is pericarditis. The cases in which endocarditis is most likely to appear cannot, we think, be accurately pointed out. Some, however, believe that the likelihood of endocarditis is in direct proportion to the severity of the joint affection.

(2) **Acute tubal nephritis.**—Next to rheumatism, Bright's disease must be considered as the most frequent cause of endocarditis, although the exact relationship between the affections is not properly understood.

(3) **The acute specific fevers,** including pneumonia, diphtheria, measles, and puerperal fever, appear to give rise to endocarditis, usually of a severe form.

(4) **Extension of inflammation.**—Extension of the inflammation, either from the pericardium or from the heart muscle, will induce inflammation of the lining membrane of the heart.

(5) **Syphilis.**—Endocarditis occasionally arises in the course of syphilis.

The male sex, middle life, debilitated conditions, especially if connected with chronic alcoholism and the presence of an already diseased valve, predispose to the severer variety of endocarditis. Of the less severe variety, those conditions which tend to produce rheumatism may be said to predispose.

Pathology and morbid anatomy.—Inflammation almost always attacks the endocardium of the valves of the heart, and not, except in rare cases, the lining membrane elsewhere. Its effects are chiefly manifested at a little distance from the edges of the cusps, and mostly on the surface of the valve which is subjected to friction or to the force of the blood-stream, for example, on the ventricular surface of the aortic valves, and at the situation where they lie in contact when closed, in fact, chiefly on their lunated edges from the corpora Arantii to their attach-

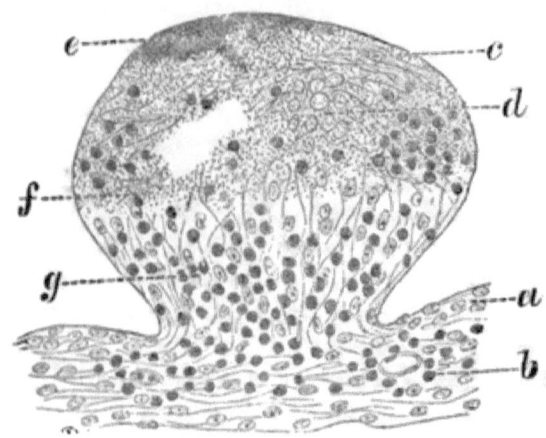

Fig. 48. Section of a nodule in malignant endocarditis (Ziegler).

a. Endocardium.

b. Subendocardial fibrous tissue partly infiltrated with leucocytes.

c. Infiltrated leucocytes.

d. Upper part of the growth consisting of fibrous and granular coagula.

e. Colourless denucleated protoplasmic masses.

f. Finely granular substances, ? micrococci.

g. Zone of transition from the undestroyed infiltrated tissue to the necrosis and coagulated tissue.

ments; and on the mitral valve on its auricular surface.

The affection attacks the left side of the heart after birth, but during intra-uterine life it affects the right side, being the cause of some of the so-called congenital malformations.

The cause of the inflammation is some irritation, sometimes apparently of a specific nature.

The process of the inflammation consists in the exudation into the tissue of the endocardium of a material which speedily undergoes organisation into a low form of fibrous connective tissue in which are many small cells.

The endothelium probably takes part in the inflammation and proliferates. The result of these changes is that the endocardium is thickened and raised into minute elevations, often covered with endothelium in a roughened condition if it be proliferated, or devoid of it altogether. These elevations form what are called *vegetations* or *granulations* upon the valves. In consequence of their roughened surface, they form points of friction to the blood as it passes over them, and the result is that fibrin may be deposited to a greater or less extent upon them. Thus, in a well-developed condition, small delicate papillæ of a red or greyish-red colour, or harder, thick, and more coarsely granular wart-like processes, of which the most superficial layers consist of deposited fibrin, are seen in the situations above indicated.

The form of endocarditis we have hitherto described is called **papillary**; there is, however, another though rarer form of the affection, but both are included under the head of acute endocarditis.

This form is **ulcerative endocarditis**. In this condition the irritation producing the inflammation is more severe, or possibly of an unusual character, with the result that the exudation, instead of undergoing organisation, increases in its cellular elements and forms pus. Irregular ulceration of the surface, with a certain accumulation of pus in the floor of the ulcer, is thus produced. The edges of the ulcer are swollen and thickened.

There appears to be strong evidence in favour of the belief that the cause of endocarditis, in some cases, is distinctly specific, but whether all endocarditis depends upon micro-organisms is not by any means certain. The relation between cases in which the cause is evidently mycotic and cases in which this is unlikely appears to be much the same as between pneumonia undoubtedly specific and pneumonia in which the specific element is far to seek. In some cases micrococci in zoogloear masses have been demonstrated by observers in the ulcerations of the endocardium. They have also been demonstrated as forming plugs, in the muscular tissue of the heart and in the blood-vessels of the kidney and spleen. The results of culture experiments and inoculations so far, cannot be considered satisfactory. If the disease is always due to micro-organisms it is questionable whether the same micro-organism is the agent in every case, as the affection arises in the course of so many different diseases, presumably produced by different kinds of micro-organisms.

The micrococci are probably introduced into the system, in some cases through open wounds, and when this method of entrance is not available, through the lungs or alimentary canal.

The results of the inflammation of the endocardium may be thus summarized:

- a. *Thickening, rigidity, and retraction of the valves.* These are the common results of the inflammatory process, from which arise valvular obstruction and incompetence.
- b. *Calcification.* Very frequently after some time the vegetations become the subject of calcification, and are felt post mortem to be of stony hardness.
- c. *Laceration of the endocardium,* either of the chordæ tendineæ or of the valve itself or of one surface of the valve only, giving rise to—
- d. *Aneurysm of the valve,* or in rare cases when laceration of the endocardium in the heart chamber takes place, the blood may be forced into the tissue of the heart itself, forming a—
- e. *False aneurysm of the heart*
- f. *Adhesions* of the valves to one another, especially of the chordæ tendineæ, producing valvular incompetence and obstruction.

A slower form of inflammation of the endocardium, resulting in thickening, and with **chronic changes in the valves,** whereby they become **unyielding and incompetent,** arises not infrequently.

The process of the inflammation is similar to that which has been described, but the steps of it are slower. It occurs at parts which have been subjected to mechanical strain.

The **remote effects of endocarditis** are to be traced to the fact that the fibrinous masses deposited upon the valves may be swept away in larger or smaller quantities by the blood current, with the production of—

Embolism and hæmorrhagic infarction in the lung, spleen, liver, and kidney;

Of *embolism and capillary apoplexy*, with or without *Softening of the brain*, and of

Gangrene of the extremities, and general pyæmia.

Chorea is believed to arise in some cases from minute embolisms in the brain, in the course of endocarditis.

Symptoms.—Cases of endocarditis may be divided into two classes, **(1) simple** and (2) **malignant or ulcerative.** The symptoms and causes differ in the two forms of the disease.

1. *Of simple endocarditis.*—Simple endocarditis constitutes by far the larger proportion of the cases. It occurs in the course of acute rheumatism, scarlet fever, and other affections. It cannot be considered an independent disease. Its symptoms, if it give rise to any, are apt to be confused with those of the disease to which it is secondary. If, however, during an attack of acute rheumatism or the like, the patient more or less suddenly exhibits all or any of the following symptoms, endocarditis may be suspected.

Pain in the præcordial region, if it has not existed there before, or an increase of pain if it has.

A **rapid, feeble pulse.**

Palpitation of the heart, the heart contracting

more **rapidly but less** forcibly than it did previously.

Dyspnœa, or an increase in the difficulty of breathing; considerable increase in **febrile disturbance.**
And the suspicion is confirmed if, *on examination of the chest,* there are found—

(a) **A more** extensive and forcible impulse than natural, or a fluttering character of **the apex-beat** on the application of the hand.

(b) Increase of the cardiac dulness due to the dilatation, which results from the valve mischief.

(c) A systolic or a præsystolic murmur, loudest at the apex, **or a double murmur in the same position,** which was not present some days or hours before, and **an accentuation of the pulmonary second sound. The aortic valves** are seldom attacked with acute **endocarditis, but when** they are there may be discovered, on auscultation, a **systolic,** a diastolic, or a double murmur, loudest over the second right interspace near the sternum and traceable up and down that bone.

2. *Of malignant or ulcerative endocarditis.*—Although simple endocarditis is generally connected with rheumatism, the malignant variety is much **less** common in those who have been previously attacked with that disease than in some other conditions. Thus in only 24 out of 209 cases collected by Osler,[*] did the affection arise during the progress of acute or subacute rheumatism. It is generally associated with **the** specific fevers. A considerable proportion of those who suffer from the disease in the malignant form are the subjects of old aortic valve disease. The symptoms of this affection are very various, and differ **considerably in the various conditions in which it** arises. Thus when arising in acute rheumatism, there may be marked head symptoms, delirium, high fever, and coma, and when in other diseases there may be all the symptoms of general blood-poisoning.

[*] 'Gulstonian Lectures,' 1885.

The affection may closely resemble typhoid or typhus fever, cerebro-spinal meningitis or hæmorrhagic smallpox. Osler strongly urges a protopathic origin of the disease in some instances. In forty-five of his cases he could find no history of any of the affections likely to produce endocarditis. Many were of a most fatal type, and no less than ten died within a week of the onset of the attack. These cases began with rigor and headache, vomiting and high fever, and sometimes there was early delirium and unconsciousness. The symptoms resembled typhus or typhoid fever or pyæmia.

Diagnosis.—From the account of the symptoms it will have been evident that the diagnosis of endocarditis, except the simple form arising in the course of an attack of rheumatic fever, is difficult and often impossible. If the malignant form occurs during rheumatic fever, however, as the heart is constantly examined, should an endocardial murmur arise, and the symptoms become greatly aggravated, and should delirium or the typhoid state supervene, the diagnosis of endocarditis will probably be made.

Again, if high fever of an irregular type arise and persist in a patient who suffers from aortic valve disease, especially if there be evidence of embolism in the lung, spleen, or kidney, malignant endocarditis is probable. But if either the simple or the severe variety of the affection arise in the course of the specific fevers, including pneumonia and diphtheria, it is very probable that if, on the one hand, the affection is slight, it will not give sufficiently characteristic symptoms to suggest a systematic examination of the heart, and that it will in consequence be overlooked, and that if, on the other hand, it is severe, its symptoms will be looked upon as those of the original disease, and endocarditis will be undiagnosed. Again, it may be mistaken for pyæmia, to which, indeed, it may give rise.

Chronic endocarditis may go on and remain unsuspected for years, the changes which damage the valves progressing so slowly that little or no inconvenience is caused to the patient, and the disease of the valve is often revealed by an accidental examination of the chest for life assurance or what not.

Terminations.—1. **Complete Recovery** is possible if the valves are not much damaged, but if possible is very rare. A method of recovery by cicatrization of the involved tissue has been described.

2. **Incomplete Recovery.**—The affection usually gives rise to chronic valvular diseases and their consequences. If the heart undergo compensative hypertrophy the patients may live for many years in comparative comfort.

3. **Death.**—*a.* From the affection itself. This is rare, and scarcely ever occurs in rheumatic endocarditis; if from the infective form, death may arise from *cardiac palsy; œdema of the lungs; asthenia from fever; softening of the brain;* or *pyæmia.*

b. From valvular diseases and their consequences: *embolism; cerebral apoplexy; gangrene of the extremities; chorea.*

Prognosis.—Although death is rare from uncomplicated endocarditis, the prognosis is always grave, since on the one hand we have to expect valvular diseases, and all their grave sequelæ, and also embolism into the various organs, especially into the brain (causing hemiplegia), and on the other there is the chance that the inflammation may extend into the muscle of the heart (myocarditis).

In the form known as ulcerative endocarditis, if it be diagnosed, the prognosis is especially grave, as general blood-poisoning is almost certain to follow.

Treatment.—At present there is no treatment which appears to have any influence in stopping or rendering less severe the course of endocarditis.

Anti-rheumatic remedies may be tried in rheumatic cases, *e. g.* carbonate of potassium or salicylate of sodium.

Anti-syphilitic drugs should be **administered in cases** with a distinct syphilitic history, **especially** in the form of large doses of iodide of potassium.

If the action of the heart is rapid and feeble give **cardiac stimulants,** such as digitalis, especially if there is any cyanosis and dropsy.

If œdema of the lung threaten, it is useful to **bleed.**

Local bleeding does not seem to be of much use unless **pain** is present; cold applied to the præcordial region is not recommended; possibly a blister would have greater effect.

Stimulants are advisable in any case in which the heart seems to be labouring much.

V. Myocarditis [inflammation of the muscular substance of the heart]

Etiology.—In this affection the muscular fibres of the heart may undergo nearly all the changes produced by inflammation.

i. Either there is a general implication of the muscle, leading to its *softening* and resulting, it may be, in its *breaking down and disintegration in patches*, leaving behind *cicatrices of fibroid tissue.*

ii. *Abscess of the heart* may arise ; or

iii. As an effect of a more chronic change, *fibroid induration* of the whole organ may remain.

The causes which produce myocarditis are as follows :

1. **Acute rheumatism,** either arising with endo- and pericarditis or else independent of one or both of these affections. In this form of the inflammation, it is commonest to find comparatively small scars of fibroid tissue, resulting from very localised patches of inflammation, but there may be a general change, and the heart-fibres may be found post mortem to be **swollen,** softened, greyish, or yellowish, and under the

microscope to have lost their striations, or there may be a so-called cardiac abscess.

2. **From extension of inflammation in endo- or pericarditis.**—When this is the case the inflammation is not as a rule so severe as when it arises independently. Valvular diseases also appear not infrequently to set up the inflammation of the muscle. Fibroid induration may arise in such cases.

3. **Pyæmia, septicæmia, typhus and scarlet fevers, smallpox** and the like may produce cardiac abscess, as may also embolism of the coronary arteries, especially in cases of gangrene of the lung.

4. **Syphilis** appears to set up the chronic form of the affection, which results in fibroid induration of the heart.

Post-mortem appearances.—Although rarer than endocarditis, the inflammation of the muscular tissue of the heart occurs more frequently than is ordinarily believed. Its seat is most often the left ventricle near the apex; but it also attacks other parts, such as the septum, and also the papillary muscles.

Effects.—i. The early stages of myocarditis are seldom observed, but if so, the condition of softening and flabbiness, and the other appearances above described, are found.

ii. Fibroid indurations, of varying size; and when spreading over large tracts of tissue, the blood-pressure may distend and bulge them out, forming—

iii. Cardiac aneurysm, in size varying from that of a hazel-nut to that of a hen's egg and upwards.

iv. Ossification of the dilated tissue may ensue or the aneurysms may become filled with laminated clot.

v. Abscesses may be found which may (*a*) open into the pericardium, producing pericarditis; (*b*) cause perforation of the septum; or (*c*) lead to acute aneurysm by the blood burrowing into an abscess cavity. It is said that an abscess may become encapsuled and dried up.

vi. The heart is nearly always dilated.

Symptoms.—Of necessity of the vaguest possible character. If during the course of acute rheumatism symptoms **of an acute character** pointing to a **weakening of the heart** occur, with or without symptoms of embolism, *e. g.* enlargement of the spleen, blood in the urine, rigors, or fever, myocarditis may be suspected, if there be **no physical signs of peri- or endocarditis.**

In other cases, when aneurysm or contractions of the heart have occurred, there is much cardiac inability, and the heart contracts with difficulty, the apex-beat is hardly to be felt, and the pulse is small, irregular, and intermittent, and there is much systemic venous congestion, with cyanosis and dropsy. But these symptoms occur in almost every chronic disease of the heart, and so are not diagnostic.

As for the other grave results of myocarditis, *e. g.* abscess and perforation, they are, as far as we know, undiagnosible.

Treatment resolves itself, as in the case of endocarditis, into a treatment of symptoms as they arise.

VI. Fatty Degeneration of the Heart

There are three varieties of degeneration of the heart into which fatty change enters to a greater or less extent. They are as follows:—(1) *Fatty infiltration of, or between the muscular fibres.* (2) *Fatty metamorphosis of the muscular fibres.* (3) *Brown atrophy of the heart-fibres.*

(1) In **fatty** infiltration of the heart there is a collection of fat upon the outside of the organ, especially marked and thick in the neighbourhood of the auriculo-ventricular grooves, and along the course of the coronary arteries, and at the apex and margins

of the heart. When this layer of fat, which is present to a greater or less extent in the normal heart, becomes very well developed, the fatty material tends to insinuate itself between the muscular fibres beneath, and in exaggerated cases so presses upon them that in the end they may atrophy. This is also accelerated by pressure upon the blood-vessels, cutting off a due supply of blood, and so stopping the normal oxygenation. The changed processes which take place in the fibres are those which are described below in the next section.

Fatty infiltration occurs in the **obese,** in whom it may be generally stated that the fat-forming elements of their food are in excess, *e. g.* the fatty, saccharine, and amylaceous elements, as well as in **old age,** where the activity of the circulation is diminished, without what is ordinarily considered disease of necessity accompanying it.

It also accompanies **chronic alcoholism** and **cancer,** with or without fatty metamorphosis.

(2) **In fatty metamorphosis** a more serious change takes place from the first. The muscular fibres themselves undergo true fatty degeneration, and their protoplasm is **converted** into fat, not as in infiltration, when the fatty material is deposited between the fibres. The change which occurs in the fibres may be local, partial, or diffuse, and if more diffuse the degree of the degeneration in different parts may vary.

(*a*) When **local,** the parts chiefly liable to be affected are the columnæ carneæ, the musculi papillares, and the superficial fibres beneath the endocardial lining.

(*b*) When **diffuse,** the change is seldom found so **far** advanced, except in those cases where a diffuse degeneration follows a local degeneration of the parts above mentioned as liable to be attacked.

The appearance of a part, the subject of this intimate fatty change is paler than natural. It is brownish yellowish or mottled in the parts affected, but may be simply of a duller red than that of a healthy organ. It is always softer and more pliable, and the tissue is more opaque. When the degeneration is diffuse the heart appears to be more flabby than is ordinarily the case, indeed, the aspect presented by a heart in this condition removed from the body a few hours after death, is suggestive of the considerable post-mortem changes observed when an autopsy has been long delayed.

The microscopic appearances of muscular fibres which are undergoing fatty degeneration vary according to the stage at which they are examined. The earliest change is an indistinctness of the transverse striations, with an appearance of granules within the protoplasm arranged longitudinally, transversely, or, it may be, irregularly in the fibre. Next, the granules increase in number, run together in parts to form small droplets of oil, the transverse striations disappear, and finally, little is left of the affected fibre but granular and fatty débris, the whole of the muscular protoplasm having entirely broken down and disappeared. In very late stages the scanty endomysium is involved in the change and nothing remains of the muscle; it is simply replaced by fat.

Causes:
A. **Local**: (1) *Insufficient blood-supply to the heart muscle* from atheroma and other affections of the coronary arteries, as in Bright's disease, or from insufficient supply in consequence of aortic valve disease, or in calcification of the arteries in old age.

(2) *From pressure upon the fibres* or upon the blood-vessels, as in fatty infiltration, or as a result of pericarditis.

B. **General**: (1) *From diminution in the quantity of the blood or in deterioration of its quality, e. g.* in anæmia after severe hæmorrhage, in chlorosis, or in fevers or in phosphorus poisoning, whereby the general oxygen-carrying power of the coloured blood-corpuscles is interfered with.

(2) *From interference with the activity of the circulation,* as in old age. Dilatation of the heart may be supposed to cause fatty changes in its walls. This is specially the case with dilatation subsequent to hypertrophy, whereby the increased muscular tissue degenerates.

(3) *From chronic diminution in the due oxygenation of the blood,* as in chronic lung disease.

(3) **Brown atrophy.**—The so-called brown atrophy of the heart is a condition in which there is a deposit of pigment within the heart-fibres. It is nearly always combined with more or less fatty infiltration. It occurs in chronic diseases of the heart and lungs and also in old age. It is a change only to be certainly diagnosed with the microscope.

Symptoms.—If fatty change arises, which is simply infiltration, produced by over-feeding and want of exercise, the activity of the organ may be very little interfered with, but, on the other hand, the condition may give rise or increase the tendency to:

Shortness of breath.
Palpitation.
Faintings.

But even if these symptoms are not marked, the affection is serious, since there is always the danger of fatty metamorphosis following it.

Fatty metamorphosis gives rise to the symptoms of those abnormalities of the heart, which are likely to come from degeneration of its muscle, viz. dilatation and atrophy. If the degeneration be local, and due to local causes, then, of course, the dilatation, &c., occurs only locally. The most usual symptoms, however, are:

Feeble apex-beat, which is more or less displaced.
Feeble and slow pulse, possibly irregular and intermittent.
A tendency to fainting.

Shortness of breath, possibly the so-called Cheyne-Stokes' respiration (p. 86).

Some œdema of the extremities and elsewhere, with blueness and coldness of the hands and feet.

Angina pectoris.

These symptoms may be due to, or associated with, the affections giving rise to the degeneration of the muscle, *e.g.* pericarditis, endocarditis, and valvular disease, and if these are present it is practically impossible to extend our diagnosis further. If, however, they are absent and the above symptoms arise, it becomes possible to arrive at a diagnosis of structural change.

Treatment.—A careful dieting will often diminish obesity. Dieting therefore, moderate exercise in the open air, and a course of mineral-water treatment at one of the German spas, of which it appears that Carlsbad or Marienbad is the best, should be recommended.

In the more serious cases, great care must be taken lest death result in one of the fainting attacks. The best treatment is that recommended under the head of dilatation, viz. rest, change of air, absence from worry, good nutritious food, and tonics. Digitalis, however, is believed to be dangerous in cases where there is marked fatty degeneration; it may produce, so it is said, rupture of the heart, and is distinctly contra-indicated.

VII. Degenerations of the Heart other than Fatty

Degenerations of the muscular structure occur in the heart occasionally, but they are chiefly interesting from a pathological than from a clinical standpoint.

They are—(*a*) Amyloid degeneration.
          (*b*) Carcinoma.
          (*c*) Tubercle.

(*a*) **Amyloid or lardaceous** disease may attack the heart and especially the right ventricle.

(*b*) **Carcinoma** is very rarely found in the tissue of the heart. When present the varieties are the medullary and the melanotic. It may occur in nodules or it may be diffuse. It may occur as part of a general cancerous infection or may extend inwards from the mediastinum. It is said that the cancerous nodules may sprout inwards into the cavities of the organ.

(*c*) **Tubercle.**—It is very uncommon to find the heart affected with tubercle. Caseous nodules resulting from pericarditis may, however, occur.

Symptoms.—Any degeneration of the muscular walls of the heart will of necessity interfere with its action. A diagnosis, however, of the special condition of degeneration, except when there is abundant evidence of the disease elsewhere, is practically impossible.

Treatment.—Symptomatic only.

VIII. Parasites

The *Cysticercus cellulosæ* has been found in the heart, so has also the *Echinococcus*.

IX. Rupture of the Heart

It may be taken as certain that no heart perfectly healthy in structure ever ruptures. The rupture of a heart even greatly degenerated is very rare.

Any degeneration of the heart tissue, especially when the walls are weakened in places, predisposes to rupture. Of these, the most common is **fatty degeneration**, but **myocarditis**, especially if causing aneurysm of the heart, is also a probable cause.

The rupture may arise from any strain put upon the organ. It may be single or double. The outside rent is regular and smooth, and is nearly always of the left ventricle.

Sudden death is the usual consequence, but there have been cases in which death has been postponed for several hours. The latter appears to be the case when the rupture at first does not pass through the whole thickness of the ventricular wall.

CHAPTER XI

VALVULAR DISEASES OF THE HEART

When inflammation of the endocardium or other disease attacks either the auriculo-ventricular or the semilunar valves of the heart, the valves tend to become abnormal in two ways, firstly, by their imperfect closure they are no longer able to prevent the backward flow or regurgitation of the blood, and secondly, by their rigidity they offer a distinct obstacle to the onward current. Thus, affections of the valves of the heart are spoken of as either **regurgitant** or **obstructive**. Regurgitation of the blood is due to incompetence of the valves. Obstruction by rigid valves is generally accompanied by narrowing or stenosis of the valvular orifice. At each valvular orifice we have two distinct abnormalities to consider, namely, incompetence of the valve with regurgitation of blood through it, and rigidity of the valve with stenosis of the orifice. Each abnormality produces marked effects upon the heart and circulation, which are more or less characteristic of the lesion; regurgitation and obstruction, however, frequently exist together at the same orifice.

The orifices of the left side of the heart are more frequently affected during extra-uterine life, as their tissue is most likely to be attacked by disease. The orifices of the right side are seldom found to be abnormal except in the cases of the so-called congenital diseases of the heart.

In considering the diseases of the valves and their results upon the heart and the circulation, attention must first of all be given to those of the aortic and mitral valves, as these are the most frequent and the most important. Under the head of aortic disease are included obstructive and regurgitant disease, and the same applies to mitral disease.

A. Aortic Disease

Causes.—The diseases which attack the aortic valves, and impair their functions, are chiefly of two kinds viz. **atheroma** and **endocarditis**. The former of these is a chronic change often associated with gout or syphilis, and is by far the more common. Endocarditis is much more likely to attack the mitral than the aortic valves; still, either in a subacute or chronic form it may affect the latter in cases of recurring rheumatism.

The aortic valves are also occasionally rendered incompetent by rupture from being subjected to a sudden severe strain. Whether this accident is likely to occur unless the tissue of the valves has been rendered inelastic by previous disease is certainly doubtful.

The conditions *predisposing* to aortic disease may be thus enumerated. Age: Since aortic diseases are usually due to the atheroma, they more frequently occur in the old, than in the young. Sex: Men are more often attacked than women. Occupation: Laborious occupations, and those in which there is much risk of exposure to extremes of heat and cold, predispose to aortic disease. Condition of life: Any circumstances which tend to set up gout must also be considered to predispose to the affections under consideration; as does also syphilis.

Post-mortem appearances.—It is rare to find the aortic valves in a state producing obstruction to the blood-stream, without finding that they are at the same time incompetent to close the orifice. The reverse of this is not, however, the case, as there can be marked incompetence without obstruction. The condition of the valves post mortem varies considerably, from a mere rim of reddish vegetations at a

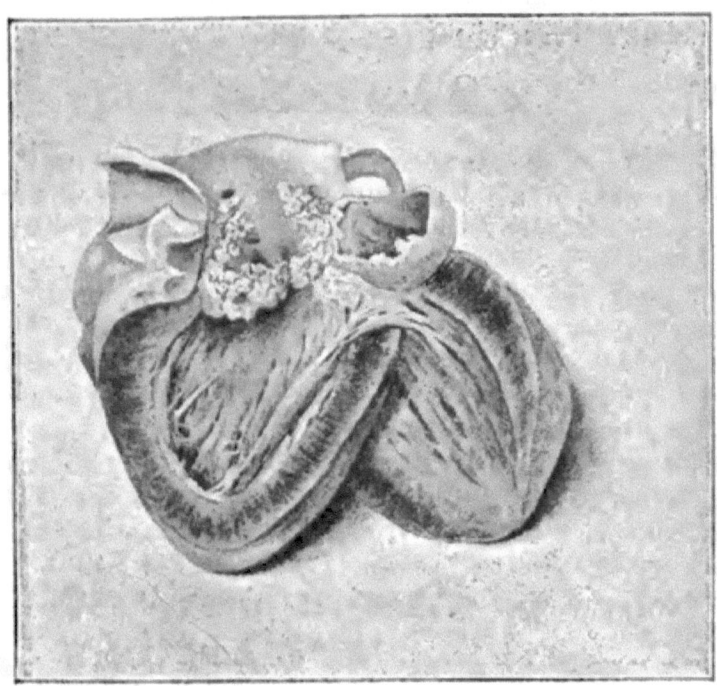

Fig. 49.—A heart from a case of ulcerative endocarditis in which the patient had numerous embolic aneurysms. The heart and aneurysms are preserved in the Museum of St Bartholomew's Hospital. (From a drawing by T. Godart.)

little distance from the free edges of the cusps in a recent case of acute endocarditis, to marked shrinking, thickening, shortening, laceration, perforation or detachment from their insertion.

Sometimes the valves have almost disappeared, leaving in their stead hard, cartilaginous, or calcified masses (Fig. 49). Again, two of the cusps may be entirely adherent, or even all three, forming a cone-like projection into the aorta; this is especially the case when the cusps adhere towards their corpora arantii.

It is rarer for obstruction to be due to **actual constriction of the aortic orifice.**

The effect of the rigidity of the valve is **to constrict the orifice to a variable extent, sometimes so markedly that it will not admit the forefinger.**

Effect upon the heart.—1. *Of aortic incompetence.*—
The primary effect is *dilatation of the left ventricle*, often to an enormous extent, so that the cavity is as large as a good sized orange; closely following upon this, is compensative *hypertrophy of the left ventricle* to an equally great extent, so that the walls may be an inch in thickness. The septum ventriculorum is pushed towards the right cavity. In addition to this, there is more or less *general hypertrophy and dilatation* of the other chambers of the heart.

At a late stage the muscle of the heart may be seen to have undergone fatty degeneration, and the mitral valve is also frequently affected, being incompetent only when the dilatation of the ventricle has passed a certain limit, or incompetent as well as rigid if the process which originated in the aortic valve has extended to it.

(2) *Of aortic obstruction.*—In those infrequent cases in which there is pure obstruction of the aortic orifice without incompetence of the valve, the left ventricle may not undergo dilatation (Fig. 50), but, in order that the blood should be able to pass the obstruction at the aortic orifice, the walls become hypertrophied. Thus a case of simple hypertrophy of the left ventricle may arise, together with more or less implication of the other chambers of the heart. Simple hypertrophy, even if it occurs, is very unlikely to persist for long without dilatation, as it is almost impossible for obstruction to exist for any length of time unaccompanied by incompetence.

Symptoms and signs.—Even those who suffer from most extensive disease of the aortic valves as a rule experience little inconvenience, because the valve lesions which tend, in both incompetence and obstruction, to diminish the blood in the arterial system and to increase that in the venous system are compensated for by the hypertrophy of the heart.

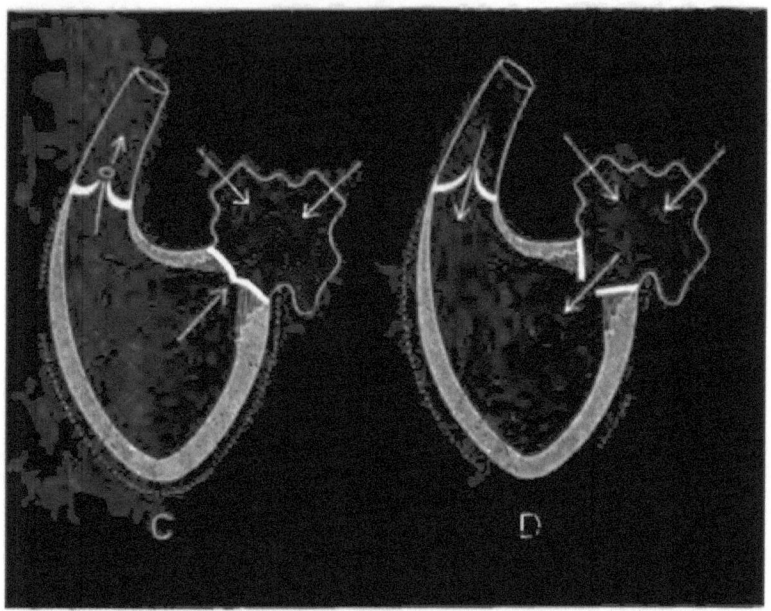

Fig. 50.—Diagrams showing the effect of aortic valve disease, C of obstruction, and D of regurgitation, upon the left ventricle. In C, as the orifice is narrowed, the ventricle hypertrophies to expel the blood with greater force through the obstructed orifice. In D, as during diastole there are two streams of blood into the ventricle, one onward from the auricle and the other backward from the aorta; the cavity dilates and the walls hypertrophy.

There are, however, three periods in the history of those who suffer from aortic valve disease which are recognisable—

1. The first is *before hypertrophy of the left ventricle* has taken place. In either form of the affection:

(*a*) The left ventricle is dilated and the circulation retarded.

(*b*) The arterial system contains less blood than it should, the arteries are unfilled, as it is said, in the one case, because part of the blood flows back into the ventricle, and in the other because less blood passes out of the ventricle at each cardiac contraction, but in ordinary cases from both of these reasons. In consequence of this the pulse is small and jerky.

(*c*) The venous system contains more blood than it should, and there is venous obstruction, both systemic and pulmonary, evidenced by blueness of the lips and face, and dropsy, together with a tendency to dilatation of the right side of the heart and congestion of the lungs and internal organs.

This stage, if it can be made out in a case of recent aortic disease, is luckily short, or the circulation would soon come to a stand-still. It speedily passes into the second stage.

2. The second stage is established when *hypertrophy of the heart sufficient to compensate for the dilatation has arisen.*

(*a*) The circulation becomes more rapid and stronger. The pulse is stronger, although jerky, and the venous system, both systemic and pulmonary, is relieved.

(*b*) Internal congestions disappear, and the right heart, if remaining a little dilated, is also hypertrophied.

(*c*) The condition of the patient is scarcely to be distinguished from one in perfect health, unless, as very often happens, the hypertrophy surpasses the dilatation, in other words, when there is too much hypertrophy for the required purpose. Under these circumstances—

Palpitation is present, but it is much more objective than subjective.

Cardiac pain, sometimes of the nature of *angina*

pectoris (p. 67), running across the chest and down the inside of the left arm, is often present.

Congestion of the cerebral circulation, shown by —Dizziness, headache, muscæ volitantes, and possibly apoplexy, as symptoms of the hypertrophy. Attacks of **cardiac asthma** may arise. If stenosis of the orifice be more pronounced than incompetence, the patients are pale, and may suffer from syncope and symptoms of cerebral anæmia.

This stage may last for many years, the patients being able to follow their occupations, even if laborious, and enjoying good general health.

It should also be remembered that many cases of aortic derangement arise from a very chronic disease of the valve, so that the first stage in them is hardly ever seen, and the diagnosis of heart disease is often not made, simply because nothing arises to suggest a physical examination of the chest.

3. The third stage arises after a variable time; in it the *hypertrophy is no longer compensative.*

The new condition is due to either (i) **fatty degeneration** of the muscle, with an accession of dilatation; or (ii) an increase in the valve-lesion from prolonged and extensive endocarditis or atheroma; or it may be due to (iii) mitral incompetence, arising either from an extension of the primary disease or from increased dilatation of the left ventricle depending upon it.

In this condition the events happen in the sequence of those which may be found in the first stage, viz. diminishing arterial and increasing venous fulness with all their chain of symptoms.

Internal congestions of the liver, kidneys, brain, and more especially of the lungs; *cyanosis; dropsy.*

Death may arise from (i) œdema of the lungs; (ii) embolism of the middle cerebral artery; or (iii) hæmorrhage into the brain.

Physical Signs.—1. *Of aortic incompetence:*
On inspection.—*a.* The apex-beat is displaced

downwards and outwards even to the seventh or eighth rib or to the mid-axillary line.

b. There is bulging of the præcordia.

c. There is a very heaving impulse, diffused and forcible, with evident pulsation of the carotids, and of those arteries, such as the temporals, which are exposed to view, in which, too, there is elongation as well as dilatation.

d. Capillary pulsation, which may be demonstrated by making some lines upon the skin of the forehead with a blunt-pointed instrument such as a penholder, and the marks are seen to become alternately pale and red.

On palpation.—*a.* The impulse is felt to be extremely forcible and diffused.

b. There is a diastolic thrill over the second right interspace close to the sternum.

On percussion.—The dulness on percussion indicates extensive hypertrophy of the left ventricle; it is increased downwards and outwards.

On auscultation.—A diastolic murmur of varying intensity and pitch is heard loudest at the second right interspace, traceable down the sternum. It is often heard and with greatest intensity over the ensiform cartilage, or even more frequently about the fourth left interspace close to the sternum.

The loudness of the murmur may be very great, and often it is to be heard all over the front of the chest and in the axilla and behind; it is sometimes heard by the patient himself. It can frequently be traced into the larger arteries, and sometimes into those of medium size, and into the radials.

It is often accompanied by a systolic murmur which is "not always due to obstruction at the mouth of the aorta, but it is probably sometimes of a mere relative constriction of the orifice connected with positive dilatation of the ascending aorta." (Gee.)

It should be distinctly remembered that a diastolic murmur *varies greatly in loudness and intensity*, and if the heart contracts feebly it may disappear, as often happens just preceding death. In certain cases it cannot be heard unless the patient lies down or sits forward, and is nearly always increased in intensity by exertion.

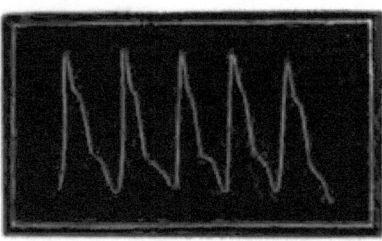

Fig. 51.—Pulse tracing in a case of aortic incompetence (amplified). The patient from whom it was taken frequently suffered from angina pectoris.

Fig. 52.—Pulse tracing in a case of aortic incompetence (amplified). In this tracing the predicrotic and dicrotic notches are well marked.

The **pulse** of aortic incompetence is very characteristic. It is known as the *water-hammer* pulse, or *pulse of unfilled arteries*, and by other names. It is jerky, sudden, and ill sustained. The rise to the maximum of tension is abrupt, but the fall in tension is sudden, and in the interval between the beats the arteries contain little blood. The characters of the pulse are well shown in a sphygmographic tracing.

There is a sudden and considerable rise of the **lever**, due to the projection of the blood by the hypertrophied left ventricle, followed by a sudden fall as the blood passes out of the arteries, both backwards into the heart and onwards into the capillaries. The dicrotic **wave is not well marked.** The characters of the pulse are exaggerated by raising the patient's arm above his head.

2. *Of aortic obstruction.*—When obstruction exists without regurgitation, *which is very seldom, if ever the case,* there is evidence of less hypertrophy of the heart; there is a systolic thrill at the second right interspace, and a systolic murmur heard loudest in the same situation, traceable into the arteries.

As a rule, however, the two affections are found together, and the so-called see-saw murmurs are loud and rasping, and audible all over the præcordial region.

B. Mitral Disease

Causes.—Mitral lesions are nearly always due to acute endocarditis, at any rate if they are primary; lesions secondary to aortic disease may be due to atheroma, and occasionally, if arising from an abnormal condition of the musculi papillares, may be a consequence of myocarditis. The mitral valve may become incompetent from dilatation of the cavity of the left ventricle.

Of predisposing causes.—Mitral disease is much more common among the young than is aortic disease, as in them the conditions from which endocarditis arises occur most frequently. It occurs in nearly equal numbers in the two sexes. Occupation, except in tending to produce rheumatism, does not seem to exercise any special influence.

Pathology and morbid anatomy.—In the earliest condition, when immediately following endocarditis, the incompetence probably arises from a mere inability of the cusps to approximate closely; later on, however, the cusps are thickened and contracted, sometimes almost calcareous, or the chordæ tendineæ are shortened, ruptured, or adherent to the wall of the cavity. In stenosis, the valve cusps are shortened and thickened, the chordæ tendineæ are adherent in part, or the lower edges of the valves are adherent, forming a constricted, funnel-shaped orifice, sometimes so small as hardly to admit the tip of a finger, or forming a narrow chink, " button hole," which is all that remains of the communication between the chambers. Sometimes both obstruction and regurgitation are due to masses of hard concretions resulting from vegetations, and sometimes these are more or less movable in the blood-stream.

Effects upon the heart and vessels.—i. *Of mitral incompetence.*—The right ventricle is greatly hypertrophied and dilated, and the walls may be as thick as those of the left.

The pulmonary artery and veins are dilated, and the walls are probably thickened.

The left auricle is considerably dilated and hypertrophied, and the right auricle to a lesser extent.

The left ventricle is dilated and hypertrophied to a moderate degree.

ii. *Of mitral obstruction.*—The effect of stenosis or obstruction is the same as in the case of incompetence, but—

    a. There is greater dilatation and hypertrophy of the left auricle and pulmonary veins.

    b. The left ventricle is not hypertrophied or dilated, indeed, the walls may be thinner than natural, and its cavity may be contracted.

Symptoms and course.—As in the case of aortic disease, so in mitral, the stages which can be made out are three in number, and represent the periods when (i) the dilatation of the heart begins, and is accompanied by symptoms denoting disturbance of the equilibrium of the circulation. (ii) When the circulation has been righted by hypertrophy of the previously dilated heart. (iii) When the hypertrophy again gives way to the dilatation, and is no longer compensatory.

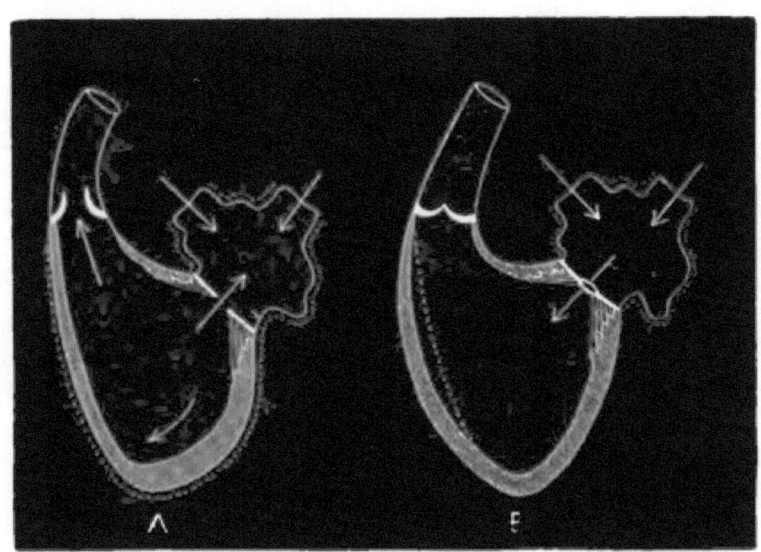

Fig. 53.—Diagram showing the effect of mitral valve diseases, A regurgitant, and B obstructive, upon the left auricle and ventricle. In A, during the systole of the ventricle, there are two streams of blood into the left auricle, one backward from the ventricle and the other onward through the pulmonary veins, and so its cavity dilates and afterwards hypertrophies; the left ventricle afterwards dilates and hypertrophies. In B the left auricle dilates and then hypertrophies in consequence of the obstruction to the exit of blood from the auricle. The left ventricle contracts rather than dilates.

First period.—When the left ventricle contracts, it discharges too little blood into the aorta, either because part of its contents regurgitates into the auricle, or because it contains too little blood from the obstruction of the flow from the auricle. In consequence of this—

a. The arterial system contains too little blood,

and the pulse is of small volume; the patient may be pale.

b. The left auricle is dilated, either because of the meeting within it of two streams of blood, backwards from the ventricle and forwards from the lungs, or because it is not emptied at each contraction.

c. The pulmonary veins are engorged and dilated, the pulmonary circulation is obstructed and congested, giving rise to the various symptoms of pulmonary hyperæmia.

d. The pulmonary artery is dilated, as is also the right side of the heart.

e. The systemic veins become engorged with blood, with congestions of the liver, kidneys, and brain, and, in the case of mitral incompetence, the left ventricle may dilate.

This derangement of the circulation gives rise to *dyspnœa*, greatly increased on any exertion; *small, feeble pulse; blueness and coldness of the extremities, blueness of the lips and tongue; œdema of the feet and legs; diminution of the urine*, which is thick, high coloured, and contains urates and probably albumen.

Second period.—The improvement of all the above symptoms begins when hypertrophy arises. This is particularly marked in the right ventricle and in the left auricle.

By means of the hypertrophy of the walls of these chambers the circulation, both pulmonary and systemic, is righted to a very considerable extent, but the former, even under these conditions, is generally overfilled, although there is no absolute obstruction.

The walls of the right auricle, too, are thickened, and in the case of mitral incompetence, those of the left ventricle also undergo some hypertrophy.

In cases of mitral disease the duration of this period is never so long as in aortic disease, neither is it so uninterrupted; indeed, there is nearly always *some shortness of breath, particularly on exertion*, and in cases of children *languor* and *apathy*, and a *more or less disinclination to engage in games with other children*. There is also a *tendency to bronchitis* from the over-full condition of the pulmonary circulation; and bronchitis when it arises tends to put the circulation still further out of gear.

Third period.—The disarrangement arising from the hypertrophy becoming non-compensative may be due to—

i. Fatty degeneration of the heart.
ii. Over-exertion of body and mind.
iii. Intercurrent affections, especially bronchitis.
iv. Extension of the original lesion.

The condition of affairs sketched as occurring during the first period becomes re-established.

The arterial system becomes less and less filled, the pulse is soft, small, irregular or intermittent. The veins are engorged with blood, and there is more or less **cyanosis and dropsy**, both of the extremities and also of the **serous membranes**.

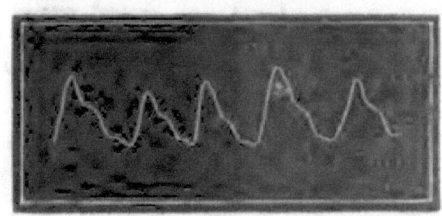

Fig. 54.—Pulse tracing from a case of mitral incompetence (amplified). It is irregular in force, frequency, and volume.

The urine is diminished, it may be to a few ounces per diem, is high coloured, and contains urates and albumen, and possibly blood and casts.

The **liver is enlarged and tender**, and there may be with this some jaundice, and all the signs and symptoms of portal obstruction, including enlarged spleen, ascites, dilated veins over the abdomen, and hæmorrhoids with gastric and intestinal catarrh. In females, menstrual derangements occur.

Signs of tricuspid regurgitation are very possibly present (p. 358).

Languor, drowsiness by day and restlessness by night. The chief distress which is ever present is increasing **dyspnœa**, and there comes a period when the patient is able to do no more than sit up in or by the side of the bed; his lips and face are bluish, his eyes are staring, his breathing is rapid, is sleepless by night; has little appetite, and irregular or constipated bowels; his legs and thighs are swollen and **tense**, it may be discharging in places from the skin **having** burst; he passes little urine. On listening to his chest sonorous and sibilant râles are heard throughout, and it is clear that there is much congestion, passing on to œdema of the lung.

It is evident that such a state cannot be prolonged; **it is either relieved by treatment** or the patient dies speedily.

Death may be **due to œdema of the lung**, to **embolism and** infarcts into internal organs, or **to some of the** intercurrent maladies indicated above.

Physical Signs.—1. *Mitral regurgitation:*

On inspection.—(*a*) The apex-beat is found to be displaced downwards and outwards, and is (*b*) more diffused and forcible than natural. It is also felt in the epigastrium. In the third period it is diffused but is not forcible.

On palpation.—It is stated that a systolic thrill is sometimes present.

On percussion.—The percussion dulness is increased chiefly in width, and is also displaced outwards and downwards to correspond with the altered position of the apex-beat.

On auscultation.—There is a loud blowing or rasping systolic murmur, heard loudest over the situation of the apex-beat, traceable to the left lateral region and heard at the angle of the scapula behind.

It is often heard all over the præcordial region.

Its character is nearly always blowing or rasping, but it may be distinctly musical.

The pulmonary **second** sound is usually **accentuated.**

2. *Mitral stenosis or obstruction:*

On inspection.—The apex-beat is seldom much displaced **but it is** forcible, and is often felt distinctly in the epigastrium and to the right of the sternum.

On palpation.—A very characteristic thrill or purring tremor, presystolic in **time, is felt.**

On auscultation.—A prolonged presystolic murmur audible at the apex-beat, or just above it, is heard; as a rule it is very localised and not traceable into the axilla, or heard behind.

The accompanying first sound is usually loud but shorter than natural, and the two natural sounds of **the heart** are often extremely like a double second sound. **The second** sound at the apex is *sometimes* **absent,** but seldom for any length **of** time.

As regurgitation through the mitral valve is often combined with stenosis the systolic murmur of **regurgitation** is frequently heard with the presystolic murmur, and may be present even when that is absent.

The **time of** the murmur of stenosis also varies, sometimes being immediately after the second sound, and at **other** times just before the first, **in** both cases of course being post-diastolic.

The murmur is generally heard **best** when the patient **lies down,** indeed, is often **heard only** under that condition.

Its variation from day to **day, and even from hour** to hour, is sometimes very remarkable.

It sometimes disappears altogether for a shorter **or** longer time.

Its characters are peculiar. It is harsh and might be represented, according to Dr Sutton, by four r's and b (r, r, r, r, and b). When present in its most typical form it can scarcely be mistaken.

C. Tricuspid Disease

Etiology.—Disease of the tricuspid valve depending upon endocarditis or atheroma is rare. Insufficiency of the valve is usually due to dilatation of the right side of the heart arising from obstruction within the pulmonary area. Tricuspid stenosis is almost invariably accompanied by mitral stenosis, and in half the cases by aortic disease.

Physical Signs.—1. *Tricuspid incompetence:*
- n inspection.—The heart's impulse is seen in the epigastrium, as well as at the apex-beat.
- On palpation.—Occasionally a systolic thrill is to be felt in the epigastrium. There is true pulsation of the jugular veins.
- On percussion.—The percussion dulness is increased to the right.
- On auscultation.—A systolic murmur is heard, loudest over the ensiform cartilage, and slightly to the left, very localised, not traceable into the axilla, or heard at the angle of the scapula behind, and most distinct when the patient is lying down.

2. *Tricuspid obstruction:*
A presystolic thrill and murmur are present, to be detected most plainly over the ensiform cartilage.

D. Pulmonary Disease

Pulmonary disease is, in the great majority of cases, congenital. It arises from imperfect development or from intra-uterine inflammation.

1. *Pulmonary incompetence:*
It is the rarest of valvular diseases.

Symptoms.—The symptoms of the affection are those of dilatation and hypertrophy of the right heart, with very considerable **dyspnœa, and symptoms** of **general venous congestion,** cyanosis and dropsy, apparently soon followed by **tricuspid insufficiency.**

Physical Signs.—In addition to the signs of hypertrophy and dilatation of the right ventricle,

A **diastolic murmur and thrill,** are most plainly heard in the second left interspace, close to the sternum. Tricuspid regurgitation with a systolic murmur and pulsation of the jugulars speedily follow.

2. *Pulmonary obstruction:*

Symptoms.—The symptoms of the affection are those of dilatation and hypertrophy of the right heart, but even when the latter is compensative, which is not common, there is considerable **dyspnœa, blueness of the lips, tongue, and face,** with palpitation.

Physical Signs.—In addition to those of a moderate degree of hypertrophy of the right ventricle there is a systolic thrill and murmur, to be detected most plainly at the second left interspace or at the third rib close to the sternum. The murmur is traceable upwards to the top of the sternum, but is inaudible in the carotids.

Diagnosis.—The diagnosis of valvular diseases depends to a very considerable extent upon the presence and recognition of cardiac murmurs. The consideration of symptoms no doubt aids the auscultator, but an absolute diagnosis is scarcely to be made unless a murmur is discovered on the application of the stethoscope. When the murmur is heard, the other methods of physical examination furnish important data. Since the diagnosis depends to so large an extent upon the auscultatory signs, attention should specially be directed to two ways in which even these break down when **too** absolute a reliance **is** placed upon them.

The two chief **ways**, then, are, firstly, **that** valvular heart **disease** may exist without a murmur, and secondly, that cardiac murmurs may arise without the **valves** being affected.

For the **sake of** convenience **it may be** as well to **classify** the anomalous cases under two heads:

I. **Cases** in which the absence of auscultatory **signs does** not exclude the existence of valvular **heart** disease.

II. **Cases** in which cardiac murmurs are inconclusive evidence of valvular lesion.

I. **Heart disease without constant auscultatory signs.**—Grave heart lesions may be present without auscultatory signs, and the first indication of anything wrong may be an attack of angina pectoris, faintness, dyspnœa, blood-spitting, collapse, or death, but these **lesions** although possibly **v**alvular are usually fatty heart, atrophy, hypertrophy, aneurysm of the heart, spasm, dilatation, or rupture from weakness of its muscular tissue. Yet it must be recollected that the best authorities agree that—

(a) *Aortic disease without constant murmur may exist*

Thus of aortic diseases, regurgitation **is** certainly **the** commonest, even supposing that stenosis without regurgitation is possible. The most certain, and, indeed, the only sign, is the diastolic murmur. Yet Fagge states that even this murmur is sometimes difficult to detect, and that it is also sometimes very slight. For example, of forty cases of aortic regurgitation examined post-mortem at Guy's Hospital during 1870–71, twenty-six cases only were rightly diagnosed, of the remaining fourteen, eleven had been removed from the surgical wards of the hospital, **or** were **less** than seven days **in** the wards, or had no notes taken of auscultatory signs. But in three cases at least the regurgitant murmur, if it had existed, must have been slight, as it was undetected.

Stokes recognised the fact, to which **special attention** should be directed, that aortic valvular disease may be latent, and unaccompanied **by constant murmur**. The reason of this he believed to be that the silent disorganizing process may gradually advance without pain, cardiac irregularity, or other symptom which would draw the attention of the patient or of his medical man to the condition of the heart; and that very likely during this period there would be absence of physical signs as well; and he added the following very important words:

"Thus it often happens that we may with great care examine the heart, and find no evidence of disease, yet in a short time, it may be in a few days, manifest physical signs are developed which indicate not a recent and acute disease, but an extremely slow and long existing affection, yet one which had not, until the period of the second examination, arrived at the point when it was at last attended with acoustic phenomena. . . . . I have known all the signs and symptoms of permanent patency of the aortic valves to occur within a few months after the effectuation of a large insurance, and yet at the period of the medical examination, which was made by one of the best observers of this or any other country, no sign of disease of the heart existed."

Nearly all writers agree with Stokes that the gravity of the aortic lesion is not in direct proportion to the loudness of the murmurs. He went so far as to state that no one could declare that, because a murmur is very distinct, the disease is very considerable; "nor can we, on the other hand, pronounce absolutely upon the healthy state of the valves merely because we can hear no murmur;" also "decrease of murmur may coincide with increased disease." This last dictum, he, however, added, applies only to obstruction of the aortic orifice—he had never observed its truth in cases of free regurgitation.

Stokes' experience in **aortic disease** was that—

i. **Aortic stenosis** may exist without a constant murmur, and that occasionally, as the narrowing of the orifice increases, a loud murmur existing before may become fainter, or disappear entirely.

ii. **Aortic regurgitation**, when sufficiently marked to be important, is always accompanied by a murmur, which is scarcely ever single; but that a stage exists antecedent to marked regurgitation in which the murmur may be very slight (difficult to detect) or absent—premurmuric stage.

Peacock admitted the possibility of stenosis of the aortic (or other) orifice existing with little or no murmur, and this, too, although he held pure aortic stenosis to be very rare.

Sanders, in a highly important paper on "The Variation and Vanishing of Cardiac Organic Murmurs," gives as his experience that the aortic diastolic murmur never completely disappears; but that it varies greatly in intensity, so as to be sometimes inaudible to unpractised ears to which it had been previously distinctly audible, and when pulmonary râles have been present, it has been sometimes quite undistinguishable by the most attentive listener; as regards the systolic murmur, he says that he had not found the same variations in intensity of the murmur as in other kinds of valvular disease, but that he had little doubt that it would be found to present some variation.

In well-marked stenosis of the aorta, **Walshe** states that weakness of ventricular action or extreme smoothness of the constricted orifice may prevent the development of a murmur; but he considers that the aortic regurgitant murmur is constant. He adds: "I can, as a bare possibility, conceive complete destruction of the aortic valves with smooth surface to exist without murmur, but have never observed the fact." Walshe records a case in which aortic insufficiency was replaced by stenosis, "a solitary example," and a diastolic aortic murmur replaced by a systolic, and a similar case has been recorded by Gairdner.

In a chapter "On the Variation and Vanishing of Cardiac Murmurs," Balfour has called attention to the fact that "even murmurs dependent upon recognised organic lesions change and vary, and not unfrequently disappear, the lesion of course still remaining." Of **aortic regurgitation**, he says: "I do not refer to those curious cases which occasionally occur, in which, apparently from deposition of fibrine upon the diseased valves, the regurgitant murmur gradually diminishes in intensity, and sometimes entirely ceases to be audible; but to cases in which, without co-existing disease of the valve, its segments have been unable to meet from over-dilatation of the aortic orifice, the result of dilatation of the aorta itself," accompanied by a soft diastolic blowing sound, which occasionally disappears entirely under the influence of treatment. Of **aortic obstruction**, he agrees with Stokes that murmurs vary very greatly in intensity, the lesion remaining, but that they never completely vanish.

As regards the relation of the loudness of the aortic murmur to the lesion it is a very general and acknowledged belief, as before said, that the loudness of the murmur is not any evidence of the degree of obstruction. H. Davies expressed this belief in the following words:—"A small amount of obstruction and roughness is occasionally attended by a much louder murmur than one of a more extensive nature, in consequence of the contractile energy of the heart being considerably greater in the former than in the latter case. Thus, in those extreme examples of aortic or mitral contraction in which the orifice has been reduced to a mere chink, the ventricle becomes incapable of propelling any considerable amount of blood from its cavity, and the resulting murmur is observed to be weak and imperfect.'

In the examination of doubtful cases of aortic disease, then, we should not neglect the advice given by Rosenstein, which embodies many well-known points of importance in the examination of cases of supposed aortic disease. "In some cases, when the patient is in the horizontal posture, a loud murmur is audible, which completely vanishes, or becomes very indistinct, when the vertical posture is assumed. In general it seems to me that the sitting posture is most favorable for hearing these endocardial murmurs. But in dubious cases it is better to let the patient exert himself a little, and then examine him in different postures before giving one's final opinion as to the presence or absence of murmurs."

Neither Guttmann nor von Niemeyer mentions the possibility of overlooking an aortic diastolic murmur. Hayden admits the possibility of both the systolic and the diastolic murmurs disappearing, the former by extreme debility of the left ventricle, the latter from two different causes—viz. temporary restoration of valvular competency by the lodgment of a plug of fibrine in the abnormal passage, and debility of the left ventricle from fatty degeneration combined with a rigid and inelastic state of the aorta.

(b) *Mitral diseases* may also exist, without constant murmurs

Disease of the mitral valve is of such frequent occurrence that great opportunity is doubtless afforded of recording cases of alteration or disappearance of previously existing murmurs, or of the occurrence of marked mitral disease without murmur throughout its whole course. Fagge, indeed, pointed out the fact that perhaps presystolic murmurs may be overlooked from their likeness to the normal first heart-sound; but even this supposition is insufficient to explain the numerous instances of mitral stenosis in which the murmur has been absent.

The same author in his original paper collected forty-seven cases of mitral stenosis found after death, in only seven of which had the presystolic bruit been discovered. And even allowing that fifteen to twenty had to be subtracted as having proved fatal before thorough examination, yet the proportion of diagnosed to undiagnosed cases was only about one to four. Further experience confirmed his belief that the cases of mitral stenosis undiscovered from absence of presystolic bruits during life are very numerous. He added that, even if a presystolic murmur has once been detected, it may for a time or altogether disappear. In the later stage of the disease, when the heart is beating quickly and irregularly, "it is almost always absent." It alters with illness, disappearing or returning with muscular exertion, with alteration of position, sitting up or standing, with running upstairs, and with similar circumstances.

Of mitral regurgitation, Fagge stated that the systolic murmur, loudest at the apex, and audible at the angle of the scapula behind, is certainly present in comparatively few of the cases. Nearly all the authors agree in the main with Fagge.

Peacock has pointed out that a mitral regurgitant murmur may entirely disappear or be obscured when the heart is embarrassed, or feeble in action, as by an attack of bronchitis; and also that in cases of advanced mitral regurgitant disease, when the heart is extremely irregular, the murmur may occur only occasionally.

Balfour states that mitral stenosis is more frequently associated with entire absence of murmur, with an irregularly intermittent systolic murmur, or with a permanent mitral regurgitant murmur, than with a presystolic murmur. He believes that certain regurgitant murmurs, from dilatation of the parts, may be entirely replaced by a perfectly normal sound, under appropriate treatment.

Walshe is somewhat guarded in what he says about the murmur of mitral regurgitation, " once established, it is, as a rule, permanent; but when probably dynamic, as in chorea, epilepsy, &c., it may momentarily, as well as permanently, disappear. If the cause of non-occlusion, even though it be organic, act intermittently, the murmur will be present at some moments, absent at others. This curious peculiarity I once observed in an adult male, in whom a body about the size of a large pea was suspended by a thread-like peduncle from the larger division of the valve." This is, however, equivalent to stating that mitral regurgitation, even when it has been due to organic change, has not always a constant murmur, or, perhaps, the loudness of the murmur may not be sufficiently great to allow of its detection. Walshe agrees with the general opinion, that the presystolic apex murmur is not infrequently wanting where mitral constriction is found after death. He has known this murmur to come and go, from day to day, in a case where the mitral orifice was greatly contracted and rigid, probably from varying force of the heart's action.

Andrew's summary of the variations, &c., of the direct or mitral presystolic murmur is as follows:— "The murmur varies in loudness, length, and pitch, may disappear entirely or even be replaced by a systolic apex murmur."

Hayden acknowledges the frequent absence of presystolic murmur in mitral stenosis.

Sanders' dictum touching the mitral systolic murmur is highly important:

" I have observed this murmur, when dependent on organic lesion of the valves, present very marked variations in loudness and distinctness, and sometimes disappear altogether for considerable periods.

". . . . I do not refer to the well-known obscuration of cardiac murmurs by the sound of pulmonary râles, or their difficulty of detection when the heart's action is tumultuous or very irregular . . . . but to . . . . variations of murmur when the circulation is in vigour, and do not refer to the disappearance of murmurs, which often takes place near the fatal termination of a case of heart disease, or in conditions of great debility."

The same author refers to the well-known changeability of the presystolic mitral bruit.

Barth and Roger, so long ago as 1842, pointed out that the absence of a murmur did not prove the healthiness of the auriculo-ventricular orifices.

(c) and (d) *Pulmonary and tricuspid valve disease*

Pulmonary valve disease is so very rare that it is unlikely that, amid the doubts surrounding the diagnosis, many observations have been taken of the possibility of the disease existing without murmur, or of the murmur or murmurs altering or disappearing, and we have been unable to find any remarks on the subject in any of the standard works on cardiac affections. The same remark applies to cases of tricuspid disease. In fact, diseases of the right side of the heart continue to be clinical and pathological curiosities. Sanders, however, infers the occasional absence of presystolic murmur in cases of tricuspid obstruction, from the fact that a considerable number of cases of tricuspid stenosis are found post mortem, which have been undiagnosed during life; and of the organic tricuspid systolic, he infers its alteration and disappearance from the fact that the murmurs due to dilatation of the tricuspid orifice in distended right ventricle are in the habit of undergoing considerable variation.

II. Cases in which **cardiac murmurs are inconclusive evidence of valvular lesions.**

The murmurs which occur independently of valvular disease are the following:

1. *Anæmic or hæmic murmurs.*—These murmurs have been already discussed (p. 110).

2. *Murmurs from dilatation of the orifice or of the ventricle.*—This kind of murmur, Gairdner says, occurs from regurgitation through the mitral or tricuspid orifices, due to the inability of the valve segments to properly close the orifices. They have been recognised by many observers, and the murmur which so often is heard in cases of palpitation may be of this nature.

In an exceedingly interesting paper on the theory of dilatation of the heart as a cause of cardiac hæmic murmurs, &c., read before the British Medical Association in August, 1882, by Balfour, many facts were brought forward to support the idea that all chlorotic and anæmic murmurs are due to dilatation. This author pointed out that the theory adopted by Hope, Bellingham, Potain, and others, as to the seat of the functional anæmic murmur at the aortic orifice, is untenable, because, firstly, it is not propagated along the aorta, or in the carotid; secondly, because its position of greatest intensity is over the pulmonary artery; and, again, that the idea that the pulmonary artery itself is the seat of the murmur is untenable, as is also Parrol's theory that the murmur arises at the tricuspid orifice. Balfour believes that the murmur may be heard over any or all of these valves. His idea of the effect of anæmia and chlorosis upon the heart is as follows :—Firstly, there is abnormal friction between the spanæmic blood and the venous walls, causing a venous hum, and the production of fluid veins at various points. Secondly, there is accentuation of the pulmonary second sound, due to increase of the intra-pulmonary pressure from obstruction depending upon loss of tone and contractile force of the cardiac muscles. Thirdly, the heart dilates and hypertrophies, and as the heart dilates the mitral and tricuspid valves become incompetent and give rise to murmurs, and, lastly, the abnormally large ventricular blood-waves give rise to systolic murmurs in the pulmonary and aortic areas. Balfour believes that the primary hæmic murmur is auricular in its origin, and that it is due to regurgitation through the mitral orifice into a dilated left auricle, as it corresponds closely to a murmur shown by F. v. Niemeyer to be due to mitral regurgitation; these views have been accepted by Paul Niemeyer, Gerhardt, and other modern authorities.

In the discussion which followed the reading of Balfour's paper, there seemed, as far as can be made out, a willingness to agree that, in chlorosis and anæmia, there might be dilatation murmurs; but Balfour's theory of the cause of the primary hæmic murmur did not meet with much favour. Broadbent said that he had no difficulty in accepting the pulmonary artery as the seat of the murmur, and suggested that it might arise at the bifurcation of the artery.

3. *Murmurs due to mechanical causes.*—These causes may be actual displacements of the heart, as, for example, by deformed chest, by the pressure of tumours, or by fluid in the abdomen. Instances are recorded of bellows-murmurs which have disappeared on tapping the belly for dropsy, and have returned on the reaccumulation of the dropsical fluid.

Murmurs may be produced by mere pressure of the stethoscope, by roughness or projection inwards of a rib, or by an exostosis, or by pressure of lymph outside the heart.

4. *Murmurs are said to occur from an unequal tension of the segments of the valves,* or of the sides of the pulmonary artery (Guttmann), also from a temporary or permanent affection of the musculi papillares, such as occurs sometimes in chorea from spasm (?) or in fatty degeneration from weakness of their constituent fibres (Guttmann, Gairdner).

5. A metallic echo of the cardiac sound may simulate a cardiac murmur; **this may** be produced by **an** air-filled cavity in the neighbourhood of the heart (possibly aided by pericardial adhesion).

6. *Intra-ventricular murmurs* may arise, which are said to be due to roughening of the inside of the ventricle, to the presence of clots, or to dilatation of the cavity.

7. *Murmurs which disappear have been observed in acute diseases,* as smallpox, typhoid, **chorea,** erysipelas. These **may** be due to temporary endocarditis, with recovery, or to dilatation of the heart. Some of these might come under head (4). In palpitation also a murmur is often heard; this may be either from dilatation **or from** the extreme rapidity of the blood-stream.

8. *Murmurs from congenital malformation of the heart, and from* aneurysms (pp. 377 and 389).

We have now shown that the auscultator **labours** under two disadvantages: on the one hand, he hears **many murmurs** which **are** murmurs not indicative of valvular disease; and on the other, certain grave valvular affections may have no constant murmurs. What is he to do? In many cases there is **not much** difficulty—for example, when a well-marked diastolic murmur is audible, there is little doubt in **the** diagnosis of aortic regurgitation, and again, when a presystolic murmur is heard loudest at the apex, mitral stenosis may be **with** confidence indicated, as also may, with less certainty, mitral regurgitation, if the murmur, which is heard loudest at the apex, can be **traced into the left lateral** region, and heard at the angle of the scapula behind, and is accompanied by a loud pulmonary second sound.

I. But supposing a murmur other than those above mentioned be heard, what shall prove that it indicates organic valvular **disease?**

Will the locality in which it is heard determine the question? Not with certainty, as functional murmurs may occur, according to the evidence of experienced observers, at any orifice, and in the area of one or all the valves.

Will the character of the murmur aid the auscultation? Decidedly not, as functional murmurs may be loud, and organic murmurs of the gravest import may be faint and slight.

Will the *rhythm of the murmur?* Not in all cases, as even regurgitant aortic murmurs which were temporary have been noticed. If systolic, nothing could be more confusing.

Will the condition of the heart? No, for in all cases dilatation and hypertrophy may be present.

II. Again, can it be said that, because a murmur is not heard on listening carefully to a heart, no disease of the valve exists? Clearly not, for the very gravest disease may exist, as has been shown, without any cardiac murmur, as in the case of mitral or aortic obstruction, or with faint and disguised or obscured murmurs in cases of regurgitation through the same orifices.

It may be argued that the anomalous cases are rare, and that the usual diagnosis of valvular disease is not only easy, but certain; and this no doubt is true, but it is to these doubtful cases that attention should be drawn. It not infrequently happens that, on examining a patient's chest, a cardiac murmur totally unsuspected is found, and an opinion as to whether the heart is permanently or dangerously diseased may be called for; in many cases it is almost impossible to give a prognosis, because the diagnosis is doubtful. In no case is it possible to definitely decide whether or not a patient is heart-whole on a single examination from auscultation alone, unless a loud presystolic or diastolic murmur is heard. Two examinations at an interval of a couple of weeks should be insisted upon.

Treatment. — It is exceedingly questionable whether any treatment directed to the cure of the original causes of **valve lesions is of the** slightest service, in other **words, one** cannot meet the **indicatio causalis.** The administration of iodide of potassium in cases of undoubted syphilitic **origin, and even when** the lesions are undoubtedly due to rheumatic endocarditis, may, however, be tried. As **regards** the **treatment of the affection itself** certain **general** rules can **be laid down,** which are much the same as those already enumerated for dilatation and hypertrophy of the heart, and can be divided into the treatment of the stages : i. When the dilatation is compensated for by the hypertrophy ; and ii. **When** the compensation is incomplete.

i. **When the dilatation is compensated by the hypertrophy :**

Every effort should **be made to avoid laying** any extra stress upon the heart.

Moderate but not excessive exercise; a light, **not** laborious, occupation ; as much as possible **in the** fresh air ; early **hours.**

Good nutritious diet; no excess of eating or drinking ; but a moderate use of alcohol is **not** contra-indicated.

No excitement of body or mind; no running upstairs, or to catch a train.

Tonics, such as iron, quinine, or strychnine.

Upon **such** lines the aortic lesions are usually sufficiently **treated, but** mitral diseases more often fall into the second condition when the hypertrophy is no longer compensative. Of course, as has been indicated, aortic affections in a late stage may also be complicated by the same trouble.

(ii) **When the dilatation is not or is no longer compensated by the hypertrophy :**

Absolute rest in bed should be enjoined ; the patient should be kept very quiet in a warm, airy bedroom ; should be fed chiefly upon liquid foods, milk, soups, beef-tea, **chicken** broth, arrowroot, custards, and the like ; a moderate amount of good wine (Burgundy or dry sherry), brandy, or, in some cases, gin should be allowed; the bowels should be regulated and kept freely open by Carlsbad salts, Æsculap, or similar mineral waters, or, if necessary, with cream of tartar, jalap, or scammony.

As regards drugs in mitral disease there is nothing like *digitalis*, either in the form of the tincture ($\mathrm{m}x - \mathrm{m}xv$) or of the powdered leaves given in the form of a pill. Of the other cardiac tonics or stimulants, *strophanthus* is well spoken of.

In aortic disease the tonics before mentioned may be exhibited, viz. iron, quinine, or strychnine.

Symptomatic treatment.— As regards further treatment it frequently becomes necessary to alleviate symptoms when they arise and are troublesome.

Dyspnœa.—When the dyspnœa is urgent, threatening immediate death from œdema of the lungs, it becomes necessary to abstract blood by venesection; ether as a diffusible stimulant, sinapisms to the cardiac region and to the feet, may also be tried. If dyspnœa be not so urgent diuretics and purgatives should be pressed; diaphoretics are hardly borne.

Œdema sometimes becomes a most distressing feature of the case, and continues in spite of medicinal treatment. The relief which the patient experiences when the skin gives way and allows the serum to transude almost suggests puncture of the œdematous part; but this is usually not resorted to, except in extreme cases, from the dangers of erysipelas, sloughing of the skin, and pyæmia. When it is done it appears to be better to make a considerable number of punctures with a needle rather than one or two larger ones with a knife. Smearing of the œdematous parts with vaseline or with similar preparations often relieves the sense of tension.

Blood spitting, either from congestion or hæmorrhagic infarction, needs no special treatment beyond that mentioned elsewhere. As a matter of fact, it often appears to relieve the patient.

Pain in the cardiac region may frequently be removed by a linseed-meal poultice, by a small blister, or by a belladonna plaster; small regular doses of morphia in aortic disease are of much service.

Palpitation may be relieved by a belladonna plaster, but as it often depends upon some digestive disturbance the bowels should be carefully looked after and the diet should be attended to.

CHAPTER XII

CONGENITAL DISEASES OF THE HEART

Etiology.—Malformations of the heart are due either to *arrest of development*, whereby the organ remains in much the same condition as it was in early or late fœtal life, or to *intra-uterine endo-* or *possibly myo-carditis*.

Omitting the curiosity of a congenital absence of the heart or of one of its cavities as incompatible with life, the following varieties may be mentioned:

1. *Deficiency of the pulmonary artery.*—When this condition is present the blood is supplied to the lungs by the aorta through dilated bronchial vessels.
2. *Contraction of the aorta above the opening of the ductus arteriosus.*—In this variety the aorta supplies blood to the head and upper extremities, and the pulmonary artery to the remainder of the body.
3. *Closure of the aortic orifice.*—The blood in this case passes from the left ventricle through an open septum into the right ventricle, and pulmonary artery and through the ductus arteriosus into the aorta.
4. *Imperfect septum ventriculorum*, with or without displacement to the right or left, and encroachment upon the right or left ventricle. In this case the aorta and pulmonary artery appear to arise from the same chamber, or the origins of the two vessels may be reversed.

5. *Stenosis or incompetence of the pulmonary or tricuspid valves.*—These abnormalities, and especially those of the pulmonary orifice, generally joined *with patent foramen ovale*, are the most common of the congenital malformations.

6. *Openings in the septa between the auricles or ventricles*, of greater or less magnitude, may exist separately, or combined with one or more of the malformations already described.

Rare cases of other anomalies have been recorded, for example, of *absence of the pulmonary* artery, of *supplementary chambers to the heart,* and the like.

Symptoms.—The usual symptoms of almost any variety of congenital malformation of the heart are:

Cyanosis.—Frequently of a very marked character, so that the patient has an intense bluish-black appearance. The intensity of this cyanosis *is due not only to venous congestion, but also to non-aeration of the blood* to a normal extent. The streams of oxygenated and deoxygenated blood being in the majority of cases not kept distinct. The cyanosis of the face, with bluish, thickened lips and nose, staring eyes, parrot-like tongue, cold breath, are highly characteristic.

Clubbed fingers and toes.—The terminal phalanges are thickened and blue, with broad, curved nails. The extremities are cold. The clubbing is due to a very chronic thickening and hypertrophy of the tissues from the chronic congestion.

Vitality is below normal.—The subjects of congenital disease of the heart, if they survive a few weeks or months, which is not always the case, are nearly always small in body, apathetic and lethargic, and take little interest in what is going on. They are in fact very like cold-blooded animals. Their temperature is low and is easily affected. Their sexual functions are little developed. They sometimes, but not often, suffer from *palpitation*, *syncope*, *dropsy*.

Pulse is small and feeble.

It is rare for the subjects of these cardiac deformities to reach adult age, still more rare for them to survive forty years. Death arises from intercurrent affections, to which they seem especially liable. They may die from congestion and œdema of the lung.

Physical Signs.—Although it is comparatively easy to diagnose that the heart is congenitally diseased, even from the above symptoms, but especially if the murmur has been heard over the cardiac region, it is difficult, if not impossible, even on careful physical examination to come to a certain conclusion as to the exact anomaly present.

i. Murmurs, however, systolic or presystolic, in time are heard in various situations over the præcardiac area. ii. Thrills may also be present. Sometimes with these signs there exists more or less hypertrophy of the heart, most frequently of the right side.

Treatment.—The treatment of those who are the subjects of congenital heart disease consists chiefly in preventing the occurrence of affections which are likely to further embarrass the already embarrassed circulation. Thus they should be carefully protected from cold, and they should be allowed considerable latitude as to occupation. Life should be made as easy as possible for them.

Complications, particularly those connected with the lungs, such as bronchitis, should be very carefully looked out for and treated.

CHAPTER XIII

DISEASES OF THE THORACIC VESSELS

I. Inflammation of the Aorta—Aortitis

Inflammation may attack each of the coats of the aorta, although with different degrees of frequency. It is almost always of the chronic form. Acute arteritis, as it is generally to be traced to the effect of wounds, injuries, cuts or twists of the vessel, or else to thrombus, or more rarely to embolus (particularly specific emboli), does not attack the aorta, except in the very rare possibility of extension of inflammation from neighbouring structures.

a. The tunica adventitia, or external coat, is not often inflamed. Acute inflammation is very rare. When inflammation does arise it is by extension from neighbouring parts, such as from the œsophagus or trachea, or from the inner layers of the artery itself.

b. The tunica media, or middle coat, is often implicated in the inflammation of the other coats.

c. The tunica intima, or internal coat, is frequently attacked by a very slow and chronic form of inflammation. The deeper layers are those chiefly implicated.

The chronic inflammation is known as *atheroma, endarteritis deformans,* or *chronic endarteritis.* The process consists of an exudation into the subendothelial layers of the coat, which contains the ordinary inflammatory corpuscles. The exudation appears, in the first place, as larger or smaller, pale red, semi-translucent, mucous-looking patches upon the inner surface of the artery, but really beneath the endothelium and in the subendothelial connective tissue. These patches of exudation may undergo (1) *organisation into fibroid tissue*, or may (2) *partly undergo organisation and partly fatty degeneration*, or may be the subject of (3) *complete fatty degeneration*, or (4) finally, of *calcification*, wholly or in part; with different results:

- (1) If **fibroid thickening** occurs alone, it may produce either hardened, bluish-white, cartilaginous patches, or a more diffuse thickening.
- (2) If **fatty degeneration** occurs, the patches may soften and break down, and there results in the deeper layers of the intima a collection of a soft, yellowish, fatty material, which is known as an *atheromatous abscess*, and this may be converted into an *atheromatous ulcer*, if the innermost layers of the coat be removed and the softened material be discharged into the bloodstream.
- (3) If **calcification** occur, it produces hard plates of various sizes, whitish-yellow in colour, by the deposition of lime salts in the fibrous tissue. It may be associated with fatty degeneration; and, indeed, the three processes, viz. fibroid induration, fatty degeneration, and calcification, may go on side by side. Sometimes the hardened patches, when examined under the microscope, exhibit a structure resembling that of *true bone*, and *ossification* is then said to have occurred.

In the early stages of atheroma, the process is entirely confined to the outer layers of the internal coat, but, later on, the tunica media is implicated in the process and undergoes fatty degeneration and atrophy, and the tunica adventitia becomes thickened and hardened.

The appearance of an atheromatous aorta is very characteristic.

The inner surface presents, on examination, a

number of yellowish, opaque patches, of smaller or larger size, hard and cartilaginous, or calcified, as well as the atheromatous ulcers above described. That these are not so superficial as they may appear to be on first sight, is proved by the fact that the inner layers of the intima may often be torn off, leaving behind the atheromatous patch. The openings of the coronary arteries are mostly, in such cases, surrounded by a ring of atheroma or calcification. Calcified patches are also sometimes depressed below the level of the coat and not raised.

The parts of the aorta chiefly subject to the inflammatory changes are those most exposed to strain, viz. *arch of the vessel* and *the places where branches are given off.*

Causes of atheroma.—(*a*) *Predisposing:* (1) Old age; (2) gout, rheumatism, and chronic alcoholism; (3) chronic Bright's disease; (4) laborious occupations.

(*b*) *The chief exciting cause,* as far as can be made out, is an (1) undue distension or strain of the artery, occasioned by a chronic condition of high arterial tension; and (2) syphilis.

The exact relationship of gout, rheumatism, chronic Bright's disease, and alcoholism, to the production of atheroma, probably differs somewhat in each case, but we have classed them together simply because their effects are probably all indirect.

Syphilis, however, appears to produce a chronic inflammation of the arteries, not to be distinguished from the disease we have been describing, and so is considered to be an exciting cause.

Effects of atheroma.—The immediate effect of atheroma is gradually (1) *to diminish the elasticity of the artery,* which as a consequence becomes (2) *distended and elongated.* When this distension becomes exaggerated it may result in a (3) *true fusiform aneurysm,* which is merely a dilated condition of all the coats.

If, however, the atheroma has resulted in an ulcer the blood-stream may distend the middle and outer coats at that point, the result being (4) a *sacculated and true aneurysm*, supposing the coats have at this point been hardened by the antecedent inflammation. If this has not been the case one of two things may happen, viz. (5) *rupture of the vessel*, or (6) *a dissecting aneurysm*; the latter occurs when the coats at the margins of the ulcer have not been strengthened and firmly joined together by inflammatory thickening, and as a consequence blood makes its way between the coats, separating them or splitting them up longitudinally. (7) *Embolism* of one or more arteries may occur from the washing away of the atheromatous débris into the circulation. (8) From the extension of the process from the vessel to the valves of the heart, *aortic stenosis and incompetence* may result.

Symptoms.—Inflammation of the aorta is chiefly known by its effects, such as have been mentioned in the last section. *A rigid radial, with a full hard pulse, tortuous arteries*, so well exemplified in the temporals in a *person past middle life*, especially if associated with some *hypertrophy and dilatation of the left ventricle*, would suggest atheroma of the aorta; as, however, the blood-supply to the heart is most likely deficient, hypertrophy soon becomes non-compensatory.

It should be recollected how prone those affected with atheroma are to suffer from aortic valve disease with its effects, which have been already considered.

Nothing definite can be said as to the symptoms of inflammation of the other coats of the aorta.

Treatment.—If there be well-supported evidence of syphilis it may be of service to attempt an antisyphilitic treatment, but otherwise attention has to be directed to the effects of the arterial affection.

II. Aneurysm of the Aorta

Aneurysms of the aorta are of two kinds, true and false. By the term* true aneurysm is here meant a dilatation of the vessel for a limited extent, the walls of which are formed of one or more of its coats. A false aneurysm, on the other hand, is one which in part, at any rate, is bounded and enclosed by the tissues or organs in the neighbourhood. A true aneurysm may be either cylindrical or fusiform, when a limited portion of the whole artery is subject to dilatations, or sacculated when only a portion of the circumference of the vessel is affected. In some cases all of the coats are dilated, in others one or two of them have given way. The external coat, as being the most distensile, is the one which usually gives away last. When the aneurysm becomes very large even this coat disappears, and the wall of the dilatation is then formed of the neighbouring connective tissue matted together and thickened by inflammation, and the organs and other tissues upon which it encroaches—the true aneurysm has become a false one. In the formation of an aneurysm sometimes the tunica intima and sometimes the tunica media is the first to give way. If the blood not only distends the external coat but also forces its way between the external and middle coats from the position of rupture of the latter as before mentioned, a dissecting aneurysm results.

Sacculated aneurysms are generally due to dilatation of small parts of the arterial circumference, and may, indeed, communicate with the lumen of the vessel by a small aperture. They vary very much in shape but are usually more or less globular.

Conditions under which aneurysms arise.—A. *Of the arterial coats.* The exact condition of the artery

---

* According to some authorities a true aneurysm is a circumscribed dilatation of *all* of the coats of an artery.

which gives rise to aneurysm in the majority of cases appears to be some **disease of the internal coat** which weakens it and causes it to **give way before** the ordinary pressure of **blood in the vessel.** Some observers, however, it should be remembered, believe that it **is to** primary abnormalities in the tunica media that aneurysms are due. As a matter of fact, **the tunica media** is very frequently implicated in the **affection of** the intima. Thus, (*a*) atheroma or chronic endarteritis with its effects upon the intima already **mentioned, and** which is the chief condition predisposing to aneurysm, causes also more or less atrophy or fatty degeneration **of the media**; and the same result is produced in the rarer instances of acute endarteritis; but (*b*) fatty **degeneration of** the internal **or middle coat;** or (*c*) simple senile atrophy of one or both of them may also predispose, and possibly in **very** rare cases (*d*) vaso-motor paralysis may take part in the causation.

B. *Of the patient.*—(*a*) Middle or advanced life distinctly predisposes, **as** aneurysms of the aorta hardly ever occur in **the young.** (*b*) The male sex. Aneurysms are certainly more frequent in men than in women. (*c*) The poor are more often affected than the wealthy, probably from the **laborious nature of** their occupations. (*d*) Laborious occupations. Aneurysms occur especially in soldiers, sailors, **smiths,** stokers, and porters. (*e*) **Alcoholic and** other excesses. (*f*) Aneurysms are more common among the English than among continental nations. Heredity is thought by **some** to predispose, as well as **an** aneurysmal idiosyncrasy, which is occasionally believed **to exist.**

c. *Exciting causes.*—Except in the rare cases of direct injury it is unlikely that a healthy artery **will** become aneurysmal. The question is, will a diseased artery under the ordinary conditions of blood pressure within it dilate into an aneurysm? It **would** seem that this is rarely the case, except in acute affections of the arterial walls. Nearly always there is a history of an extra strain, or a fall, or an excessive muscular effort, or perhaps a blow from which the first symptoms of the aneurysm date, and in many cases it is probable that the hypertrophy of the heart, when found associated with aneurysm of the aorta, preceded and did not arise from, the aneurysm.

Post-mortem appearances.—An aneurysm of the aorta is generally extra-pericardial, but it may be, in rare cases, within that sac, and if so, is certain to be comparatively small and liable to speedy rupture. If extra-pericardial, it may be found implicating the ascending, transverse, or descending portion of the arch of the vessel. The aneurysms are most frequently found on the convex aspect of the aorta. In size the swelling varies from that of a small nut to that of a man's head. The direction of increase in size varies with the part of the artery implicated. Thus, if the aneurysm be of the ascending part, it tends to grow forwards and to erode the sternum, the upper ribs, and their costal cartilages, chiefly of the right side; if the transverse portion of the arch be affected, the increase is upwards and forwards and to the right; and if the descending portion be affected, it tends to increase backwards and to erode the vertebræ, but may compress the left bronchus. In the much rarer cases of aneurysms being found in the concavity of the arch of the vessel, if of the ascending portion, the aneurysm tends to press upon and possibly to erode the pulmonary artery or right auricle, and if of the transverse portion, it may press upon the trachea or œsophagus.

Aneurysms may burst internally or externally, but much more often the former than the latter. They may by pressure produce erosion of the walls of any hollow viscus or tube, and may open into it. They may become adherent to any organ, so that even if their walls are thinned, as they are supported by the neighbouring structures they do not give way.

Within sacculated aneurysms there is nearly always found more or less clot, sometimes arranged in distinct layers or laminated upon the walls of the artery; it is by an increase in the amount of this laminated clot that spontaneous cure of an aneurysm takes place. Very often, however, the clot is not deposited upon the walls, but is free within the sac of the aneurysm and is frequently irregular and broken up. Laminated clot is seldom found in cylindrical and fusiform aneurysms, and clots of any kind much less often than is the case in the sacculated variety. An aneurysm should be distinguished from a general dilatation of the aorta.

The effect of an aortic aneurysm upon the heart is not constant. Sometimes there is associated with it dilatation and hypertrophy of the left ventricle, at other times fatty degeneration or atrophy of the heart.

When an aneurysm is near the place where a branch of the aorta comes off, the branch very frequently takes part in the dilatation; this is especially true of the innominate artery. In addition to the many other effects of pressure upon the thoracic contents, it must be remembered that a large aortic aneurysm may produce very great alteration in the position of the viscera.

An aorta which is aneurysmal almost invariably shows the signs of atheroma or other disease elsewhere. Sometimes two aneurysms are found. In a case we have recorded, one aneurysm about the size of a hen's egg was found quite solid and full of laminated clot, while the other was small, recent, and quite without lamination. When an aneurysm is cured by laminated clot, the clot becomes organised and the tumour tends to get smaller.

The presence of an aneurysm is a considerable source of irritation to the neighbouring serous membranes, from which acute inflammation may be set up.

Symptoms and course.—The diagnosis of an aortic aneurysm, even if it be sufficiently large to affect by pressure the neighbouring structures, is not easy, unless it tend to produce an external pulsating tumour. As this is not by any means an invariable tendency, it comes to pass that aneurysms of the aorta are sometimes discovered first of all on the post-mortem table.

The diagnosis, even in doubtful cases where the tumour is small, is however often to be made by careful attention to symptoms produced by the increase of pressure of the aneurysm upon neighbouring structures, and by a careful and thorough physical examination of the chest. In cases where the aneurysm is large, if not externally evident, the same care and attention are required to render the diagnosis certain. An aneurysm of the aorta may give rise to some or all of the following symptoms, which are mostly the effects of pressure:

Pain.—The pain experienced in cases of aneurysm of the aorta is of two kinds, arising from two different causes. The first kind is neuralgic and intermittent, being subject to acute exacerbations. It proceeds from pressure upon the intercostal and other nerves. The second kind is a constant burning, boring, and gnawing pain, which arises from pressure upon and erosion of the vertebræ or other bones. This latter pain is so agonising in some instances as to make the patients shriek out, and in a case recorded elsewhere it apparently had the effect of producing acute mania.

Dyspnœa arises from several causes, e.g. (*a*) from pressure upon the lung, trachea, bronchi, or pulmonary vessels; (*b*) from pressure upon one or other vagus. When this is the case, very severe paroxysms of difficulty of breathing occur. (*c*) From pressure upon one or other (most likely the left) recurrent laryngeal nerve.

Dysphagia which is due to pressure upon the œsophagus, or upon the vagus on either side which supplies it with nerves.

Inequality of the pupils.—This symptom, most important from a diagnostic point of view, arises from pressure upon the sympathetic nerve of one side. The first effect, or symptom of irritation, of the nerve is dilatation of the pupil on the affected side, and the later effect, or symptom of paralysis, is contraction of the pupil on the same side.

Changes in the voice, hoarseness or aphonia, arise from pressure upon the vagus or upon the recurrent laryngeal, which produces spasm or paralysis of the intrinsic laryngeal muscles of one side. If there be not actual hoarseness or aphonia, the voice is weak and deficient in timbre and may run into falsetto. All the muscles supplied by the nerves are not, however, always involved, at all events at first, but the abductors are affected before the adductors.

Cough.—Associated with pressure upon the recurrent laryngeal nerves are also to be noted stridulous breathing and brassy cough. Cough may also arise from pressure upon the trachea, bronchi, or lung. Such pressure nearly always sets up catarrh, and each paroxysm of cough is generally finished by the expectoration of a small amount of glairy mucus.

Congestion and œdema of the head, neck, chest, and upper extremities, and tortuous veins, especially marked over the chest, occur from pressure upon the veins within the thorax, generally upon the left superior innominate vein or upon the right auricle or vena cava. If one of the veins only be obstructed then the venous congestion is found only on the side affected.

Inequality of the radial pulses.—The volume of the pulse in the two radials may differ, and the pulses may not be exactly synchronous. This latter inequality is chiefly likely to arise when the aneurysm affects the root of the innominate or other large artery which springs from the arch. It is rarer for the pulse in one radial to be before the other, than that the volume of the pulse should differ. The difference in the time of the radial pulses is said to

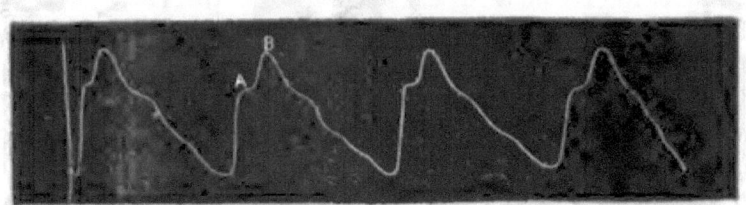

FIG. 55.—Sphygmographic tracing taken from the radial artery in case of aneurysm of the aorta (much amplified). A, the anacrotic or percussion wave; B, the pulse or tidal wave, which is higher than A, and indicates a continued rise in tension.

be one of the most important proofs of a retarded circulation. It is a distinct pause, so it is said, between the beat of the heart and the arterial pulse wave occurring at a point below the aneurysm.

Smallness of the radial pulse may be due simply to pressure upon the root of the innominate and left subclavian arteries, or to stretching or displacement of their origins by the aneurysmal tumour.

Physical Signs.—No distinct physical signs of aortic aneurysm itself are to be made out until it be of such size as to approach the chest wall. Any signs antecedent to this are those produced by its commencing pressure upon some neighbouring structure, e. g. upon the bronchus or lung, causing an alteration in the respiratory murmur.

On inspection.—A swelling is seen upon the surface of the chest, generally to the right of the sternum, involving the second and third ribs, and their corresponding interspace, or involving more or less of the upper part of the sternum, generally hemispherical or possibly nodular in shape. In the rarer cases in which the tumour appears posteriorly a swelling of **varying size**, sometimes so great as to push the scapula outwards, is found about the third dorsal vertebra.

On palpation.—Before the aneurysm comes to the surface of the chest wall and appears as a tumour it is often possible by careful palpation to discover an abnormal **pulsation** in the second or third right interspace or its neighbourhood, or in the episternal notch, or to the left of the upper portion of the sternum, or possibly behind on the left of the third dorsal vertebra. The pulsation is **systolic** in point of time, **but may be diastolic**. The pulsation may be accompanied **by a thrill, systolic or diastolic**. The amount of pulsation varies; **if the aneurysm be undergoing cure by the deposition of laminated clot it becomes very feeble**. When the **aneurysmal tumour becomes evident upon the surface of the chest, the pulsation is usually distinct**.

On percussion.—Dulness, frequently **merging into the cardiac dulness, is made out by percussion** over the upper portion of the sternum, or to the right over the second and **third ribs and the** intervening space, **or to the left of the sternum**, or posteriorly to the **left of the third dorsal vertebra**.

On auscultation.—One or both heart-sounds may be heard over the aneurysmal tumour; or a systolic or diastolic murmur, or both; or the second **sound may** be accentuated, with or without a **systolic** murmur. Sometimes **no** distinct sound or murmur **is heard**.

Terminations:

1. **In recovery**; sometimes a spontaneous cure occurs from the deposition of laminated clot.
2. **In death**.
    a. *By rupture*, almost always **internally**, into the pleura, pericardium, trachea, or œsophagus; **a few** cases have occurred into the

pulmonary artery or **vena cava**, and sometimes **externally**.

 b. By intercurrent disorders, *e.g.* by spasm of glottis, paraplegia, pericarditis, pleurisy and the like.

Diagnosis.—Aneurysm of the aorta may be simulated by—

1. **Aneurysm of the innominate artery.**—In this case the diagnosis is difficult, if not impossible. The following symptoms are *more likely to occur with innominate aneurysm*, viz. feebleness and retardation of right radial pulse, with the tumour and physical signs more **towards** the right sterno-clavicular articulation, and pressure upon the superior cava, upon the right bronchus, and upon the brachial plexus.
2. **Mediastinal tumours (p. 401).**
3. **Empyema (p. 141).**
4. **Carcinoma of the pleura.**—Carcinoma of the pleura is nearly always secondary, the pulsation is not as a rule lateral, neither are the pulses different in the two radials. In carcinoma, even if there be a systolic murmur from pressure upon the aorta, there is **no** other alteration in or addition to the heart-sounds.

Prognosis.—As recovery is rare, the prognosis is of course not favorable, but patients with aortic aneurysms, although liable to sudden death, may yet live for many years.

Treatment.—Aneurysms of the aorta within the thorax admit of few methods of treatment. The kinds of treatment which are employed in external aneurysms, by distal or proximal ligature, and by compression, are of course out of the question. There remain three methods to which we can have recourse viz. (1) Tufnell's treatment, by rest and restricted diet; (2) Treatment by drugs, particularly iodide of potassium; (3) Treatment by galvano-puncture.

1. *Tufnell's treatment.*—The suggestions as to the best method of producing consolidation of internal aneurysms, made by Tufnell some years ago, were prompted, as he tells us, by nature's own cure of certain isolated cases of aneurysm in which consolidation of the aneurysmal sac had spontaneously occurred.* The publication of many cases of the successful treatment of aortic aneurysm in which these suggestions had been put into practice, has caused this treatment, at once simple and efficient, to be generally adopted.

In the treatment of all aneurysms, both external and internal, *two chief objects are to be kept constantly in view*—firstly, the prevention of an undue strain upon the aneurysmal walls, already weak and diseased; and secondly, the bringing of the blood and the circulation into such a condition that consolidation of the contents of the aneurysm may take place. In external aneurysm the surgeon is able to use several methods of producing these ends. He may diminish the flow of blood through the aneurysm either entirely by proximal ligature, or partially by regulated compression, or by distal ligature. In internal aneurysm we have a more difficult problem, as these operations are out of the question; but the indications for treatment are still much the same, for in them the flow of blood must be diminished, and the blood-pressure must for a time be lessened. There are two chief methods for meeting these indications, namely, by (1) diminishing the action of the heart, and (2) diminishing the quantity of the blood in the vessels. Valsalva, who was the first to suggest the treatment modified by Tufnell, thought that bleeding would effectually produce the desired diminution in the quantity of blood, and so he recommended it in conjunction with starvation. By this procedure, however, according to Tufnell, the blood will be found, even if diminished in

* 'The Successful Treatment of Internal Aneurysm.'

quantity, yet deficient in fibrin-forming factors, and so in a very unsuitable condition for clotting.

Disregarding, therefore, as impracticable other ways of diminishing the blood-pressure in the aorta, and shunning bleeding as likely to interfere seriously with the second object he had in view, that is to say, the formation of laminated fibrin, Tufnell directed most of his attention to diminishing the action of the heart, and at the same time to lessening the quantity of the blood, by means of the following method. He enjoined (1) *absolute rest* in the recumbent position for a considerable period, rest such as to lay upon the heart the least possible work, and so reduce to a minimum any antecedent increased cardiac stroke, which would increase the tendency of the diseased vessel to distend. In his own words, "The recumbent position is the main point to be attended to. If this cannot be steadily maintained for a considerable length of time, all other treatment will fail. In the horizontal posture the circulation is tranquillised, and the heart's action becomes regular and slow." In addition to rest (2) *restricted diet*—the amount taken to be as nearly as possible eight ounces of fluids and ten of solids a day—as likely to diminish the absolute amount of blood, and also to make it rich in solids, and especially in fibrin-forming constituents; and lastly, certain necessary *remedial agents*, such as anodynes, aperients, &c., which are likely to aid in removing any possible disturbing elements. Such is the plan which is generally known by the name of "Tufnell's treatment."

From an experience now extending over some years, it may be affirmed of this treatment, that in *a fair proportion of cases cure results*, and in *nearly every case relief of symptoms* undoubtedly occurs if the treatment is faithfully carried out by doctor, nurse, and patient.

2. *Treatment by drugs.*—The administration of

large doses of iodide of potassium as many as ninety grains daily, for months, is strongly recommended by Dr G. W. Balfour as highly beneficial, and as likely to procure consolidation of the contents of the aneurysmal sac. Ergotine, to cause contraction of the sac, and astringents, such as tannin and lead acetate in large doses, have also been suggested as useful remedies.

3. *Treatment by operation.*—The treatment of aortic aneurysm by *galvano-puncture* has been tried with success in some cases. Thus of twenty-three cases treated in this way by Ciniselli, five were cured. The method promises the most successful results in aneurysms of medium size, when still intrathoracic, and with a small opening into the vessel.

We know of no successful case of treatment of aortic aneurysm within the thorax by *packing the sac with wire or with horsehair,* which has been done in abdominal aneurysm and in that condition appears to promise fairly.

III. Fatty Degeneration of the Aorta

This is a condition which it is necessary to separate from atheroma. In both affections there is a fatty change, but in atheroma the primary disease is inflammatory, and the fatty change is of the products of the inflammation, whilst in this the fatty degeneration itself is a primary affection of the tissue itself.

The primary fatty degeneration attacks the *tunica intima* as a rule, but may also be found in the other coats.

The affected parts may be seen as small, yellow, irregular-shaped patches on the inner surface of the aorta, more superficial than atheroma, and may be stripped off, leaving behind the other coats, generally speaking, unaffected. The first change takes place in the endothelium and subendothelial tissue-cells, in both of which fat- or oil-globules accumulate, and finally replace the protoplasm. The change next attacks the intercellular tissue, and after a time the tissue of the internal coat is, in the places affected by the process, almost entirely replaced by fat.

If the fatty tissue should soften superficially small erosions of the inner coat may result, the softened material having been by degrees swept into the blood-stream; as before mentioned, the fatty change sometimes extends more deeply, and attacks the middle coat.

IV. Rupture of the Aorta

Rupture of a healthy aorta is extremely unlikely to occur. If the coats are the subject of disease, fatty degeneration or atheroma, it is more likely, although still rare.

If the rupture be of the internal and middle coats only, the blood may make its way between the middle and outer coats, and a dissecting aneurysm is the result. In such an event death may be postponed for a time. The blood finally ruptures the external coat, and finds its way into the pericardium, pleura, or elsewhere.

Rupture is attended with violent pain and collapse.

V. Constriction of the Aorta

A condition of general constriction of the aorta is a congenital malformation. It occurs chiefly in women who are chlorotic. It is accompanied, as a general rule, by delayed growth and development. The heart in such cases is either small like that of a child's, or hypertrophied and dilated to a marked degree. During life, in addition to anæmia, there may be present, palpitation, a tendency to syncope, coldness of the extremities, a tendency to hæmorrhage, profuse or absent catamenia, tendency to gastric ulcer, and to endocarditis. There is usually a systolic murmur heard over the præcordia, and, in addition, there may be signs of hypertrophy of the left heart.

VI. Dilatation of the Aorta

A **uniform** dilatation of all the coats of the aorta, especially of the arch, is sometimes to be recognised. It is to be distinguished from **aneurysm** proper, although a very distinct line of demarcation is not always present, either anatomically or clinically. In a dilated aorta there is usually visible and palpable episternal pulsation, if the part affected be the ascending or transverse portion of the arch, together with a little impairment of resonance over the manubrium sterna and an accentuated second sound over the aorta. This condition is usually associated with hypertrophy of the left ventricle.

VII. Disease of the Pulmonary Artery

Acute arteritis is most uncommon. Chronic arteritis may occur in mitral disease, especially stenosis, even if the aorta is unaffected. This has been suggested as the cause of pulmonary emboli in such cases.

- **Dilatation** to a moderate extent occurs in obstruction of the pulmonary circulation with dilatation of the right heart.
- **Aneurysm** is exceedingly rare and is seldom diagnosed. The following signs have been noted: A *swelling*, more or less, of the left side of the sternum about the second and third ribs, with *pulsation*, some *dulness*, and *loud systolic* or *diastolic murmurs*.

VIII. Pulmonary Embolism

Pulmonary embolism, or blocking of a larger or smaller branch of the pulmonary artery by a clot, gives rise to **hæmorrhagic infarction**. It is a condition in which there is a defined and circumscribed capillary hæmorrhage into the lung-tissue, which does not produce its destruction but only displacement. It may be restricted to areas as small as single lobules. The blood is effused into the alveoli and into their walls as well as into the interalveolar tissue.

Pulmonary embolism arises either in (1) **heart disease**, particularly valvular disease, in which there is dilatation and hypertrophy of the right side of the heart. In such cases pulmonary embolism is most likely to occur when the hypertrophy is not compensative, and when there is a tendency towards the formation of coagula in the right ventricle. This condition is found most frequently in mitral lesions, and apparently more often, in proportion to the number of cases, in stenosis than in incompetence of the valve. Embolism may also occur in degeneration of the muscle and other affections of the heart, and in pulmonary emphysema and bronchitis; or (2) in **thrombosis of the systemic veins** arising from whatever cause, but not of the portal vein. In pulmonary embolism, therefore, the origin of the embolus is to be looked for in the right ventricle between the columnæ carneæ, in the apex, in the auricle, or in the peripheral veins.

Hæmorrhagic infarction.—The size of an infarct varies with the size of the branch of the artery, which is occluded, from that of a hazel-nut to that of a hen's egg, and occasionally may occupy a third or a half of a lobe of the lung. The shape is generally that of a wedge with the apex towards the interior of the lung. As regards number, there may be only one, but as a rule there are several. If small, they are found at the surface of the lung underneath the pleura; if larger, they are more internal, and towards the root of the lung, being found most frequently in the lower lobe. In appearance an infarction, when recent, is reddish black; it is hard, inelastic, and void of air. On section, it is irregular, coarse, and granulated, and when scraped, a semi-fluid, brownish-red material is found upon the knife.

Microscopically, the capillaries are found to be distended, and the air-vesicles and the interalveolar tissue are packed with blood-corpuscles.

Infarcts of longer standing are paler and yellower, the fibrin having undergone fatty degeneration, and the blood-colouring matter considerable change. Still later in the history of an infarct the fatty fibrin undergoes absorption, and the hæmatin is converted into black pigment. Finally, then, the only remains of an infarct is a blackish induration, which in some cases becomes calcareous. It is usual to find the plug in the pulmonary vessel which causes infarction, but this is not always so, and it is suggested that in those cases where no plug is found that the infarct is due to primary disease of the vessel, probably fatty degeneration, plus an increase in blood-pressure.

Instead of hæmorrhagic infarctions merely, *patches of pneumonic infiltration, small abscesses,* or *circumscribed patches of gangrene,* may occur from the blocking up of small branches of the pulmonary artery; but these are either late stages of hæmorrhagic infarction or are due to the septic character of the occluding plug.

When infarctions are superficial, the pleura covering them is generally inflamed.

Pathology.—Although the post-mortem appearances of hæmorrhagic infarctions in the lungs are as a rule easy of recognition, their pathology is by no means so certain. Cohnheim's explanation, based on experiments upon the frog, is briefly as follows:

If a plug form in a *terminal* vessel, that is to say, one from which no main branch is given off below the occlusion, the first change is, that stagnation of the blood takes place in the artery itself, in the capillaries and in the veins supplied by it. After a shorter or longer time, the blood in the veins of the neighbourhood, however, begins to flow backwards into the veins of the occluded area and then into the capillaries and into the artery itself, until the whole of the vessels are filled to repletion, and may give passage to the coloured corpuscles through their walls into the alveoli, into the alveolar walls and into the interalveolar tissue. In this way small hæmorrhages arise around the walls of the vessels which continually increase and coalesce, and a hæmorrhagic infarction is the result. If the artery be not a terminal one, but if a branch comes off below the plug, a collateral circulation may be set up and no infarction may result.

Symptoms.—The symptoms of hæmorrhagic infarct vary according to its extent and to the nature of the embolus from which it arises.

1. If the *emboli be small and take their origin from the heart*, there is usually cough, with a considerable amount of frothy and blood-stained mucous expectoration, seldom with pure blood ; pain in the side or lower part of the chest, from the consequent pleurisy ; some feverishness, the temperature being raised to 100° or 102° F.; and an increase in the difficulty of breathing. These symptoms continue for a few days and then abate.

2. If the emboli be *large, and large tracts of the lung-tissue be implicated* in the infarct from the blocking of a large branch of the pulmonary artery, there is sudden and intense dyspnœa and collapse. The patient has an air of intense anxiety or despair, his face is pale or blue, his extremities are cold, his pulse is small, rapid, and thready. If he recovers from the shock of the attack, there are, in addition, cough and copious blood-stained sputum, as in the former case. There is also evidence of circumscribed consolidation of the lung, which may be followed by pleurisy and pneumonia. There may also be signs of thrombosis of the heart, with *sudden irregularity*

of the pulse; increase of cardiac dulness and cessation of any adventitious murmur.

3. If the *emboli be derived from the systemic veins*, there may be depression of spirits, and asthenic fever, but there is as a rule no cough or characteristic sputum, and there may be indeed no objective or subjective signs whatever.

4. If the *emboli be composed of septic material*, there is still more unlikelihood of characteristic symptoms being made out. The metastatic abscesses to which they give rise may be diagnosed only if there be pains in the chest, shiverings, with cough and expectoration, such as has been described, and friction or abnormal dulness be detected on physical examination.

IX. Pulmonary Apoplexy

In pulmonary apoplexy there is a profuse and severe hæmorrhage into the parenchyma of the lung, ploughing it up and destroying it, from which a cavity in the lung-tissue may result. It is almost invariably due to erosion by disease of the larger vessels, especially of the arteries, or to laceration by injuries, blows, contusions, and penetrating, including gun-shot, wounds.

In rare cases it is due to atheromatous degeneration of the pulmonary artery, with or without aneurysmal dilatation, with final rupture.

After death, a cavity is found in the lung-tissue containing liquid and clotted blood, surrounded by the disintegrated pulmonary substance. The pleura may have been torn, and blood poured into the pleural sac.

Symptoms. — This fatality is more common among middle-aged people, and particularly in men. There is violent and rapidly fatal hæmoptysis following some injury to the chest, such as a stab or gun-shot wound, or death may arise more gradually from suffocation from the blood blocking the bronchi, and not being expectorated fast enough. Sometimes the fatal hæmorrhage is entirely internal.

X. Disease of the Intrathoracic Great Veins

Acute inflammation of the great venous trunks within the thorax is practically unknown.

Dilatation of the veins occurs in pulmonary congestion.

Pressure upon and obliteration of veins may occur within the thorax from aneurysm or tumours.

CHAPTER XIV

INTRATHORACIC TUMOURS

Intrathoracic tumours include aneurysm of the aorta, but as this has been already considered, the present chapter will be devoted to solid tumours or new growths of the mediastinum.

Conditions under which they arise.—Scarcely anything is known of the causation of mediastinal growths. They occur rather more frequently in men than in women, and chiefly from puberty to middle age. Occasionally the growths appear to commence after some injury to the thorax. Hereditary taint may be suspected in some cases.

Morbid anatomy.—The majority of the tumours are found in the anterior mediastinum, but they may occur in the posterior mediastinum also. The growths are most commonly lympho-sarcomata, but others may be met with, carcinomata next in frequency to sarcomata. Fibromata, osteomata, lipomata, and dermoid cysts occasionally occur. The growths are as commonly primary as secondary.

They appear to arise either in the connective tissue or substernal lymphatics, in the bronchial or cervical glands, in the bones, in the pericardium, or in a persistent thymus.

When there is a distinct anterior mediastinal growth, there is nearly always infiltration of the bronchial glands.

In their growth they may press upon the pericardium, heart, and lungs; may press upon, erode, and grow into the trachea; **may involve** any of the great vessels, **veins, or arteries; may** compress and ulcerate into the œsophagus; may implicate any of **the** nerves of the thorax, *e. g.* the vagi, or the recurrent laryngeals. The direction of the **growth** may be forwards, to erode **the sternum and ribs,** possibly fracturing the former, when the growth may appear on **the** front of the chest; or upwards into the neck; backwards **to** press upon the heart and **lungs;** or downwards **to** depress the diaphragm.

Pleural or pericardial effusions **are** frequently present, together or separately.

In the lungs, pericardium, and elsewhere, there are almost certainly some larger **or** smaller secondary deposits.

Symptoms.—Mediastinal growths present very many of the symptoms of aneurysm of the aorta which are due to pressure (p. 386) upon neighbouring tissues and **organs; for** example, there is *dyspnœa, pain, dysphagia, changes in the voice, œdema and cyanosis* of the upper part of the body, with *tortuous veins* in the neck and over the chest. Changes in the pupil are not frequent.

There **may be cough with blood-stained frothy sputum.**

On Physical Examination.

   On inspection.—There may **be distinct** dilatation of one side of the thorax with bulging of the sternum or its cartilage forwards, and impaired movement with displacement of the heart.

   On palpation.—The vocal fremitus is generally diminished.

   On percussion.—There is extensive irregular dulness **over the front of** the chest with increased resistance.

   On auscultation.—The breathing sounds are feeble and distant over the tumour. If the bronchus or trachea is implicated there is bronchial or tracheal breathing, with increased voice.

The heart-sounds are usually **weak and seldom** accompanied by a murmur.

Diagnosis.—When hard lumps are **found at the** back of the chest, or a tumour appears external to it, the diagnosis becomes comparatively easy if the pressure symptoms are distinct and well marked. Even if there is no external tumour, if constant pain in the chest be present, and on examination some dulness with irregular boundaries is found either in front or behind, together with bronchial or tubular breathing in abnormal positions, or absence of breath-sounds, and dilatation of the thorax on one side, added to hoarseness or aphonia, blueness or œdema of the face, neck, and upper extremities, a mediastinal tumour may be diagnosed. In the absence of true pulsation, thrills, murmurs, or changes in the heart-sounds and of changes in the pulse, the diagnosis of a growth and not of an aneurysm is probable.

But it must be remembered that such signs may be wanting, and yet an aneurysm may be present; and again, signs closely simulating those associated with aneurysm may be present, and yet an aneurysm may be absent.

For example, a mediastinal growth may press upon the aorta and cause a murmur, and a murmur is also said to be present over very vascular growths; or else pulsation of the aorta behind may cause apparent pulsation in the tumour.

If there be distinct evidence of hypertrophy of the heart, aneurysm and not a new growth is, however, probable.

Mediastinal tumours, especially if cystic, may be mistaken for pericardial effusion, but the irregular shape of the dulness, the history of the case and the absence of fever, with the marked pressure symptoms, would determine the diagnosis.

From pleural effusion, the position and extent of the dulness, the history of the case, and the pressure symptoms, ought to be sufficient to distinguish them.

From **phthisis**, much the same considerations would determine the nature of the affection.

Treatment.—The treatment is merely palliative or symptomatic. The strength of the patient must be maintained by good nutritious diet, **tonics, and** stimulants if necessary. **Symptoms** must be treated as they arise. Tracheotomy is sometimes performed for the relief of urgent dyspnœa, but without much benefit. Cysts, if diagnosed, may with distinct advantage be punctured.

INDEX

A

Abnormal apex-beat, 87
— heart-sounds, 106
— pulsations, 88
— respiratory movements, 85
Abscess, lung, 220
Action, lungs, 22
— vagus, 23, **36**
Actions **of drugs,** 25, 63
Activity, resp. centre, 25
Acts, special respiratory, 24
Acute Bright's disease, 294
Acute bronchitis, 176
— — course, 176
— — physical signs, 176
— — symptoms, 176
— lobar pneumonia, *see* Croupous
— miliary tuberculosis, 266
Added **sounds,** 101
Adherent pericardium, 304
— — definition, 304
— — pathology, **304**
— — physical signs, 304
— — symptoms, 304
Ægophony, 103, 144
After-treatment
— pleurisy, 162
— pneumonia, 229
Air-tubes, blood-supply, 19
Alar chest, 84
Albuminuria in phthisis, 276
Alcohol in phthisis, 285

Algiers, 287
Altitudes, high, in phthisis, **286**
Amyloid disease, 147
— — heart, **334**
— — kidney, 280
Amyl nitrite, 74
Amphoric resonance, 96
Analyses **of** pleural fluid, 136
Anæmic murmurs, 110
Aneurysm, aortic, 382
— — exciting causes, 383
— — p.m. appearances, **384**
— — of heart, **334**
Andrew, Dr, 248, 366
Angle of Ludwig, 4
Angina pectoris, 67
— — appearance in, 69
— — attack, 69
— — causes, 68
— — definition of, 67
— — pathology, 71
— — symptoms, 68
— — treatment, 74
— — varieties, 70
Anterior mediastinum, 9
Anthracosis pulmonum, 265
Antiseptic inhalations, 289
Anvil sound, 104
Aorta, 45
— aneurysm of, 382, 383
— arch of, 45
— constriction of, 394
— dilatation of, 395
— fatty degeneration of, 393

Aorta, inflammation of, 378
— rupture, 394
— relations, 45
Aortic aneurysm, 382
— — causes, 383
— — conditions, 382
— — **cure of,** 385
— — **drugs,** 392
— — operation, 393
— — physical signs, 388
— — **p.m.** appearances, 384
— — prognosis, 390
— — **sym**ptoms, 386
— — treatment, **390**
— — terminations, 384
— — Tufnell's, 391
— disease, **343**
— — causes, 343
— — **effects,** 345
— — p.m. appearances, 343
— — signs, 348
— — stages, 346
— — symptoms, 346
— — varieties, **345**
— — treatment, **372**
Aortitis, 378
Apex-beat, 31
— — abnormal, 80
Aphthæ, 279
Apnœa, 24
Area of cardiac dulness, 96
Arteries, diseases of, 378
— coronary, 27
— intercostal, 5
Arterial murmurs, 112
— **sounds,** 112
Ascot, 287
Aspirator, 156
Asthma, 182
— **causes,** 182
— **prognosis,** 185
— p.m. appearances, **184**
— prognosis, 185
— symptoms, **184**
— treatment, 185
— bronchial, 182
— cardiac, 70

Asthma, dolorificum, **67**
— **hay, 196**
Asphyxia, 24
— blood-pressure in, **49**
— condition of heart, **49**
— — **vessels, 50**
— symptoms, 49
Astringents in hæmoptysis, **63**
Atheroma, 380
Atrophous emphysema, **203**
Atrophy of heart, 323
— — acquired, 323
— — congenital, 323
— — etiology, 323
— — p.m. appearances, **323**
— — symptoms, 324
— — treatment, 324
— — varieties, 323
Attachment of ribs, 2
Auricle, left, 29, 33
— right, 27, 32
Auriculo-ventricular sulci, 32
— — valves, 30
Auscultation,
— immediate, 97
— mediate, 97
— of breath-sounds, 99
— of heart, 104
— of pleural **sounds, 103**
— **of voice,** 102

B

Bacillus **tu**berculosis, *see* Tubercle bacilli
Balfour, Dr G. W., 393
Barrel-shaped chest, 83
Bell-sound, 104
Bergeon, 108
Bidder's ganglia, 36
Binaural stethoscope, 97
Bismuth in diarrhœa, 284
Blood discharged at each beat, 42
— pressure, **39**
— — in asphyxia, 49
— supply air-tubes, 19
— — lungs, 19

Blowing respiration, 100
Bony thorax, **diameters of, 8**
Bordighera, **287**
Bournemouth, 287
Bowditch, Dr, 299
Breaking-down stage, 277
Breast pain, suffocative, 67
Breathing, difficulty of, *see* Dyspnœa
— bronchial, 99
— cavernous, 100
— tubular, 100
— vesicular, 100
Breath-sounds, 99
Bronchi, **11**
— dilatation, 176
— divisions of, 16
Bronchial breathing, 99, 101
— **causes of,** 101
— **spasm,** 182
Bronchiectasis, 86
— diagnosis, **189,** 237
— pathology, **187**
— physical signs, 187
— symptoms, 188
— treatment, 189
— varieties, 187
Bronchitis, acute, 176
— **causes,** 174
— **chronic,** 178
— croupous, 181
— in the newly-born, 177
— in the old, 177
— pathology, 175
— treatment, **179**
— varieties, **176,** 178
Bronchioles, 13
Bronchophony, 102
Bronchorrhœa, 181
Brunton, Dr L, 56, 72
Buchanan, Dr, 248
Burdon Sanderson, Dr, 251
Buttermilk in phthisis, 285

C

California, 288
Canary Islands, 287

Cancer, *see* **Carcinoma**
Cannes, 287
Cape of Good Hope, 288
Capillary bronchitis, **177**
Carcinoma, heart, 339
— lung, 239
— pleura, 177, 239
— pericardium, 308
Cardiograph, **35**
Cardiography, 118
Cardiac aneurysm, **334**
— cycle, **105**
— ganglia, 36
— murmurs, 107
— pain, 66
— sedatives, 44
— tonics, **44**
— variation in, 108
Cardio-pulmonary murmurs, 112
Cartilages, ribs, 2
Caseation, **256**
Caseous **pneumonia, 257**
Catarrh, acute, of bronchi, 176
— chronic, 178
— **epidemic, 193**
— **fever, 176**
— of trachea, 174
Catarrhal pneumonia, causes, 230
— — course, 234
— — diagnosis, 234
— — pathology, 230
— — physical signs, 233
— — symptoms, 233
— — terminations, 234
— — treatment, 235
Causes of asthma, 102
— cough, 54
— bronchial breathing, 100
— cracked-pot sound, 95
— hæmoptysis, 61
— heart-sounds, 104, 105
— hydrothorax, 162
— palpitation, 64
— vesicular murmur, **101**
— mitral disease, 351
— atheroma, 380

INDEX

Cavernous breathing, **100**
Cavities of heart, 27
Centre, respiratory, 23
Centric dyspnœa, 51
Cerebral symptoms, **277**
Changes in urine, **223**
Charcot's crystals, **58**
Chauveau, 107
Chest contents, 9
— contractions, **83**
— description, 1
— dilatations, **83**
— external dimensions, 8
— subnormal types, 84
— viscera, with respect to, 31
Cheyne-Stokes' breathing, **86**
Cheyne Watson, 201
Chlorodyne, 284
Chloral, 284
Church, Dr, 294
Chordæ tendineæ, 29
Chronic bronchitis, 178
— course, **179**
— **physical** signs, **179**
— prognosis, 179
— symptoms, 178
— terminations, 179
— treatment, 179
Cirrhosis of lung, *see* Interstitial pneumonia
Climatic treatment, **286**
Closure of paracentesis, **154**
Cobblers, deformities of, **4**
Cocaine, 284
Cod-liver oil, 284
Cold baths, 281
Cold, rose, 196
Collective investigations, 251
Columnæ carneæ, 29
Complications in paracentesis, 159
Concentric hypertrophy, 312
Conditions of increased resonance, 94
— — diminished, 95
— — pulse, 123, **128**

Congenital diseases of the heart, 375
— etiology, 375
— **physical signs, 377**
— symptoms, 376
— treatment, 377
Congestion, hypostatic, 219
Consumption, *see* Phthisis
Convulsions, 220
Contagion, phthisis, 250
Copper sulphate, 284
Coronary sinus, 27
— artery, 27
Cough, causes, 54
— centric, 55
— character, 56
— description of act, 54
— dry, 56
— **hoarse, 56**
— loud, **56**
— metallic, 56
— moist, 56
— nerves in, 55, 56
— nature of, 54
— paroxysmal, 56
— stomach, 55
— tone, 56
— treatment, 60
— **varieties of, 54**
— whooping, 56
— in phthisis, 273
— — pneumonia, 222
Counter-opening, 161
Cracked-pot sound, 93, 278
Crackling, 101
Creaking, 101
Crepitation, 101
Crepitant râle, 101
Crepitus, 103
Crisis in pneumonia, 224
Croupous bronchitis, 181
— — prognosis, 182
— — symptoms, 182
— — treatment, 182
— pneumonia, 211
— — causes, 213
— — complications, 226

Croupous bronchitis, course, 224
— — micro-organisms in, 215
— — nature, 213
— — pathology, 216
— — physical signs, 225
— — prognosis, 226
— — sequelæ, 220
— — stages of, 216
— — symptoms, 220
— — terminations, 225
— — treatment, 229
— — varieties, 216
Cruveilhier, 140
Cyanosis, 300, 305, 324
Cycle, cardiac, 105
Cysticercus cellulosæ, 340
Czermak, 37

D

Davos Platz, 287
Dawlish, 287
Deformities, cobblers', 4
— compositors', 4
Delirium, 223, 300
Depressing expectorants, 60
Depression of ribs, 4
Depressor nerve, 41
Description of chest, 1
Diagnosis of hæmoptysis, 62
— hæmatemesis, 63
— pulmonary disease, 359
Diameters, bony thorax, 8
— external of chest, 8
Diaphragm, 6
Diaphragmatic asthma, 186
— pleurisy, 163
— — symptoms, 164
— — terminations, 165
Diarrhœa, 276, 284
Dicrotism, 128
Diet in phthisis, 280
Dietetic treatment, 285
Difficulty in breathing, *see* Dyspnœa
Dilatation of bronchi, 186
— — varieties of, 186

Dilatation of chest, 83
— — bilateral, 83
— — unilateral, 83
— of heart, 315
— — conditions of, 316
— — course, 321
— — etiology, 315
— — physical signs, 320
— — p.m. appearances, 318
— — symptoms, 318
— — terminations, 321
— — treatment, 322
Dimensions of chest, 8
Diminished resonance, 95
Diplococcus, 215
Displacement of apex-beat, 87
Divisions of bronchi, 16
Dobell, Dr, pancreatic emulsion, 285
Donaldson, Dr, 162
Doubled heart-sounds, 106
Dropsy, 320, 323
Drugs, action, *see* Action of drugs
Dry crepitation, 101
Dudgeon's sphygmograph, 120
Duration of phthisis, 280
Dysphagia, 300, 387
Dyspnœa, 48
— causes, 51
— in pneumonia, 222
— in aneurysm, 387
— varieties, 52

E

Echinococcus, 340
Echo, metallic, 103
Effects of atheroma, 380
— mitral disease, 352
— pleurisy upon chest, 136
Elastic fibres in sputum, 58
— how to find, 58
Elasticity of lung, 16
Emaciation, 275
Embolism as cause of gangrene, 238

Emphysema, **subcutaneous,** 82
— of lung, **200**
— atrophous, 203
— etiology, 200
— expiratory, 202
— inspiratory, 202
— method of production, 202
— secondary, 203
— interlobular, 203
— **free,** 203
— symptoms, **205**
— physical signs, 207
Empyema, 137, *et seq.*
— results of, 144
— symptoms pointing to, 141
Endocarditis, 324
— diagnosis, 331
— effects of, 334
— etiology, 324
— prognosis, 332
— physical signs, **330**
— **results,** 328
— **symptoms, 329**
— **terminations, 332**
— treatment, **332**
— varieties, 327
Endocardium, 30
Engorgement of lung, 216
Epidemic influence, 295
— catarrh, 193
Epidemics of influenza, 193
— pneumonia, 213
— whooping-cough, 190
Epileptic neuralgia, 67
Ergot, 283
Etiology, emphysema, 200
— **influenza,** 194
— **hay** asthma, **196**
— **rose** cold, 196
Events synchronous with heart-sounds, 104
Examination of sputum, 58
Excentric hypertrophy, 312
Exciting causes, aneurysm, 383
Exercise, 280
Expectorants, 60
— action, 60

Expectorants, varieties, 60
Expiration, 7, 21
— ordinary, **7**
— forced, **7**
— **muscles of, 7**
Expiratory emphysema, 202
Extent of pleura, 67
External dimensions of chest, 8

F

Fagge, Dr. H., 360, **364**
Fainting, *see* Syncope
Family predisposition in phthisis, 243
Farnborough, 281
Fatty degeneration,
— **of aorta,** 393
— — heart, 335
— — — causes, 336, 337
— — — diffuse, 336
— — — local, 336
— — — symptoms, 338
— — — treatment, 339
— infiltration, 334
— metamorphosis, 336
Fever, hay, 196
— phthisis, 283
— pneumonic, 221
Fibres, elastic, in sputum, 59
Fibroid induration, 147, 235
Fibroid phthisis, 147, 235
Fissures of lung, 16
Fistula in ano, 277
Flat chest, **84**
Floating ribs, 2
Fluctuation, 90, 91
Fluid, pleural, 27
— pericardial, 18
Force of contraction of heart, 38, 40
— expiration, **7**
— inspiration, **7**
Frémissement cataire, 90
Fremitus, vocal, 88
— cavernous, 89
— friction, 89

INDEX

Fremitus, rhonchial, 89
French Riviera, 287
Frequency of heart, 38, **41**
— of pulse, 123
Fresh air, 280
Friction fremitus, 89
Fuchsine staining, **59**

G

Gairdner, Dr., 73
Ganglia, cardiac, **36**
— inhibitory, **38**
— cardio-inhibitory, 384
— accelerator, 38
Gangrene, 220, 238
— in pneumonia, 220
Gargling, 102
Gee, subnormal chest, 84
Giant-cells, 261
Granulation, grey, 260
Grey granulation, 260
Growths, pericardium, 308

H

Hæmic dyspnœa, **51**
— murmurs, 109
Hæmoptysis, 61
— causes, 61
— characteristics, 62
— diagnosis, 62
— in phthisis, 275
— — syphilis, 293
— **prognosis,** 63
— treatment, 63
— v. hæmatemesis, **63**
Hæmorrhage in relation to phthisis, 253, 279
Hæmothorax, 173
Hamilton, Dr D. J., 12
Hampshire, **287**
Hay asthma, 196
— **fever,** 196
— etiology, 196
— symptoms, 198
— treatment, 198

Hazeline, 283
Health resorts, 286
— — home, 287
Heart, 26
— cavities, 27
— congenital diseases **of, 375**
— position, 31
— hypertrophy, 309
— impulse, 34
— innervation, 35
— movements, 34
— palpitation, 64
— sounds, **104**
Heat dyspnœa, **51**
Heberden, **67, 72**
Hectic fever, 275
Hensley's, Dr, method of operation, 158, 161
Hepatisation, 217
— grey, 218
— red, 217
Hereditary predisposition to phthisis, 243
Herpes, 223
High altitudes, 286
High tensional pulse, 127
Hoarseness, 276, 284
Holden, 8
Home health resorts, 287
Hope Dr, 78
Hoppe-Seyler, **136**
Hopefulness, 275
Hydro-pericardium, 305
— causes, 305
— p.m. appearances, 306
— physical signs, 306
— symptoms, 306
— treatment, 306
Hydropneumothorax, 165
Hydrops pericardii, 305
Hydrothorax, 162
— causes, 162
— symptoms, 162
— treatment, 163
Hyères, 287
Hyperpnœa, 48
Hyperpyrexia, 229

Hypertrophy of heart,
— — amount, 311
— — causes, 309
— — changes produced by, 313
— — concentric, 312
— — excentric, 312
— — diagnosis, 314
— — morbid anatomy, 311
— — nature, 313
— — pathology, 311
— — physical signs, 313
— — simple, 312
— — symptoms, 313
— — treatment, 315
— — varieties, 312
Hypostatic congestion, 219

I

Idiopathic pericarditis, 295
Ilfracombe, 287
Impulse of heart, 27
Incompetence, pulmonary, 358
Increased resistance, 96
— resonance, 94
Inequality of pupils, 387
— — pulses, 387
Inflammation of aorta, 378
— — causes, 380
— — effects of, 380
— — symptoms, 381
— — treatment, 381
— of pleura, 130
Infrequent pulse, 124
Influenza, 193
— etiology, 194
— prognosis, 195
— symptoms, 194
— treatment, 195
Inhalations, antiseptic, 289
Inhibition of heart, 37
Inhibitory ganglia, 38
Innervation of the heart, 35
Insomnia, 284
Inspection, method of, 80
— of apex-beat, 36
— — muscular system, 82

Inspection of apex-beat, nutrition, 82
— — respiratory movement, 85
— — shape and size of chest, 83
— — skin and superficial parts, 80
Inspiration, 6, 21
— muscles of, 6, 7
Inspiratory emphysema, 202
Intercostal arteries, 5
— muscles, 5
— nerves, 5
— spaces, 6
Interlobular emphysema, 203
Intermittent venous murmurs, 114
Interstitial pneumonia, causes, 235
— — in phthisis, 229
— — pathology, 236
— — physical signs, 237
— — symptoms, 236
— — treatment, 237
Intestinal ulceration, 229
Intrathoracic tumours, 401
— — conditions, 401
— — diagnosis, 403
— — morbid anatomy, 401
— — symptoms, 402
— — signs, 402
— — treatment, 404
— veins, disease of, 400

J

Jaundice, 223

K

Koch, 242, 251
— bacillus tuberculosis, 242
Koumiss, 285

L

Laennec, 68, 287
Landmarks of chest, 4
Laryngitis, 280

INDEX

Latham, 67, 69
Layers of pleura, 17
Lebert, 251
Ligamentum latum pulmonis, 17
Lobes of lung, 16
Low-tensioned pulse, 127
Ludwig, angle of, 4
Lungs, structure of, 14
— description of, 14
— apex, 15
— borders, **19**
— lobes, 16
— fissures, 16
— appearance in health, **16**
— elasticity, 16
— roots of, 18
— blood-supply, 19
— nerve-supply, 20
— lymph-supply, 20
— action in health, 22
— movements, 22
— position to chest **wall, 31**
— emphysema, 40
Lymphatics of lung, 20
Lysis in pneumonia, 224

M

Mackenzie, Sir M., 196
Madeira, 287
Mahomed's sphygmograph, 120
Malignant endocarditis, 327, 330
Manubrium sterni, 3
Marey, **35**
— sphygmograph, **120**
— tambour, 121
Marienbad, 339
Mediastinum, 9
— anterior, 9
— posterior, 9
— superior, 10
Mediate percussion, 91
Mentone, 287
Metallic echo, 103, 169
— resonance, 96
— tinkling, 169

Methods of physical examination, 80
— of production, emphysema, 200
— — mitral disease, 351
Micro-organisms, 58
— pneumonia, 214
— phthisis, 59, 270
— whooping-cough, 190
Midriff, 6
Miliary tuberculosis, 266, 279
Minute structure, lungs, 19
Mitral valve, 30
— disease, 351
— causes, **351**
— pathology, **351**
— effect, 352
— symptoms, 352
— physical signs, **356**
Mother-cells, 261
Motor centres of **heart, 37**
Movement of ribs, 4
Mucous membrane of bronchi, 12
— — of trachea, 11
— râles, 102
Murmurs, anæmic, 110
— aneurysmal, 389
— arterial, 112
— artificial, 113
— cardiac, 107
— hæmic, 109
— inorganic, 110
— local arterial, 112
— musical, 108
— pericardial, 111
— pleuro-pericardial, 111
— **pneumo**-pericardial, 111
— **venous,** 113
Murrell, Dr W., 74
Muscles, external intercostal, 5
— **internal intercostal,** 5
— **of expiration,** 7
— of inspiration, 6
Musical murmurs, 108
Myeloid cells, 261
Myocarditis, 299, 333

Myocarditis, effects, 334
— etiology, 333
— pathology, 334
— symptoms, 335
— treatment, 335
Myoidema, 275

N

Nerves, depressor, 40
— intercostal, 20
— of lungs, 20
Neuralgia, epileptic, 67
New growths, 308
New Zealand, 286
Nice, 287
Niemeyer, F. von, 147, 245, 254, 267, 301
Niemeyer, Paul, 369
Night sweats, 275, 283
Nitrite of amyl, 74
Nitroglycerine, 74
Nœud vital, 23
Non-valvular murmurs, 108
Normal percussion of chest, 92
Note, percussion, 92

O

Obstruction, aortic, 342, 345
— pulmonary 359
Œdema of chest wall, 82
— of the lung, 147
Openings of the chest, upper, 4
— — lower, 4
Opium in angina pectoris, 75
Origin of phthisis, 243
Orthopnœa, 49
Osteal note, 42
Oxygen, standard of air, 24
— effect of diminution of, 24

P

Palpation, method of, 88
— of cardiac region, 90
— of cavernous fremitus, 89
— of friction fremitus, 89
Palpation of fluctuation, 90
— of rhonchial fremitus, 89
— use of, 88
— of vocal fremitus, 88
Palpitation of heart, 64
— — causes, 65
— — symptoms, 65
— — treatment, 66
Papillary endocarditis, 327
Paracentesis pericardii, 304
— thoracis, 151
— indications for, 151
— operation described, 153, 155
— position of puncture, 153
— varieties of operation, 154, *et seq.*
Parasites in heart, 340
Parietal pleura, 17
Pathology, bronchiectasis, 181
— pleurisy, 133
— pneumonia, 216
— whooping-cough, 191
— mitral disease, 351
Pectoriloquy, 102
Penzance, 287
Percussion, method of, 91
— cause of note, 93
— character of, 92
— immediate, 91
— mediate, 91
— normal of chest, 93
— special sounds, 95
— resistance, 96
Peri-bronchitis, 259
— cardial thrills, 91
— — murmurs, 111
Pericarditis, 294
— causes, 294
— course, 299
— dry variety, 296
— effect of, 298
— symptoms, 300
— physical signs, 301
— prognosis, 303
— tapping in, 304
— treatment, 303
— varieties, 296

Pericardium, 26
— new growths of, 307
Peripheral resistance, 43
Phthisis pulmonalis, 242
— — sequel to pneumonia, 220
— — age and sex in, 247
— — apex, why affected, 245
— — bad hygienic conditions, 249
— — bronchial ulceration in, 259
— — catarrhal patches in, 256
— — cavities, 255
— — climate and locality, 248
— — contagiousness, 251
— — course, 272
— — cretaceous masses, 256
— — croupous patches in, 257
— — exposure, 253
— — family predisposition in, 243
— — fibroid growths, 255, 258
— — hereditary predisposition, 243
— — hæmorrhage as a cause, 253
— — infection, 250
— — occupation in, 249
— — over-lactation in, 250
— — pathology, 266
— — peri-bronchitis in, 259
— — pigment in, 256
— — pleura in, 255
— — pleurisy in, 253
— — pneumonia as a cause, 253
— — race, influence of, 248
— — softened patches, 262
— — struma in, 244
— — symptoms, 272
— — tubercle of pleura in, 256
— — — bacillus in, 253
— — — true, 260
— — wetness of soil as a cause, 249
Physical examination, methods of, 80

Physical signs (*see* under each heading)
— — bronchitis, 129
— — carcinoma, 241
— — bronchiectasis, 189
— — croupous pneumonia, 225
— — interstitial pneumonia, 237
— — catarrhal pneumonia, 233
— — whooping-cough, 225
— — emphysema, 207
— — mitral disease, 356
— — tricuspid disease, 358
— — pulmonary, 359
— — congenital, 377
— — pleurisy, 141
Pigeon-breast, 84
Pitch of percussion-note, 92
Plessor, 91
Plethoric cases, 228
Pleura, parietal, 17
— carcinoma of, 173
— extent of, 17
— fluid of, 18
— tubercle, 173
— visceral, 17
— diseases of, 130
Pleural fluid, 136
Pleurisy, 130
— diaphragmatic, 161
— dry, 134
— purulent, 137
— serous, 135
— sero-fibrinous, 134
— treatment, 150
— varieties, 133
— v. enlarged spleen, 149
— v. hepatic enlargement, 149
— v. pneumonia, 148
Pleuro-pericardial murmurs, 111
Pleximeter, 91
Pneumatometry, 117
Pneumococcus, 213
Pneumogastric, *see* Vagus
Pneumograph, 114
Pneumonia, caseous, 257
— catarrhal, 210

Pneumonia, croupous, 210
— interstitial, 210, 235
— lobar, 211
— varieties, 210
Pneumonic fever, 211
Pneumo-pericardial murmurs, 111
Pneumo-pericardium, 301
— causes, 307
— physical signs, 307
— prognosis, 307
— symptoms, 307
Pneumothorax, 165
— causes, 165
— course, 168
— diagnosis, **169**
— gases in, **167**
— perforation in, 167
— prognosis, 168
— **signs**, 169
— symptoms, **168**
— termination, **168**
— treatment, 172
Pointing of **empyema, 141, 146**
Pond's sphygmograph, 120
Porritt, Mr N., 9, 154, 159
Posterior mediastinum, 9
Position of heart, **31**
— — lungs, 31
— — valves, 34
Potain's **syphon-trocar,** 156
Powell, Dr D., **17,** 151, 166
Precautions in paracentesis, 159
Predisposition to phthisis, 243
Prognosis (*see* under each disease)
Prognostics in pleurisy, 149
Pterygoid chest, 84
Puerile breathing, 99
Pulmonary artery, disease, **395**
— incompetence, 358
— **symptoms, 358**
— **signs,** 354
— embolism, 395
— — causes of, 395
— — pathology, 397
— — symptoms, 398

Pulmonary infarction, **396**
— apoplexy, 399
— — symptoms, **399**
— **disease,** 358
— obstruction, 359
— pulse 122, definition, 122
— — dicrotism, **128**
— — duration, **128**
— — frequency, **123**
— — rapidity, 128
— — rhythm, 124
— — **tension, 126**
— — **volume, 126**
— water-hammer, 350
Pulsations, abnormal, 88
Pulsus bigeminus, 125
— paradoxus, 126, 301
— trigeminus, 125
Purring tremor, 90
Pyæmia, 295
Pyothorax, 137

R

Râles, 101
Red-currant jelly sputum, 240
Reduplication of heart-sounds, 106
Regurgitation, aortic, 342
— mitral, 351
— pulmonary, 358
— tricuspid, 358
Relations of aorta between **pleurisy and** phthisis, 147
— — — between phthisis and hæmoptysis, 253, 275
Remak's ganglia, 36
Resorts, health, 186, *et seq*.
Respiration, abnormal, 85
— valves, 23, 25
— movements, 21
— special acts, 24
Respiratory dyspnœa, 57
Resistance, percussion, 46
Result of empyema, 145
Retraction of the chest after pleurisy, 145

Rheumatism, 294, 325
Rhonchial fremitus, 87
Rhonchus, 101
Rhythm of pulse, 124
Ribs, attachments, 2
— floating, 2
— movement, 4
Rickety chest, **84**
Rigor in pneumonia, 220
Rindfleisch, 269
Rose cold, 196
Rotation of ribs, 4
Rupture of aorta, **394**
— — heart, **34**
— — causes, **340**

S

San Remo, 287
Scherer, 136
Sea voyage, 286
Sedatives, cardiac, 44
— vascular, 44
Semilunar valves, **31**
Sequelæ of croupous pneumonia, 220
Secondary emphysema, 203
Shortness of breath, 49
Sibilus, 101
Sibson, **Dr, 294**
Siderosis pulmonum, 265
Sidmouth, 287
Sinus, **coronary, 28**
Sleep **sweats, 275**
Softening, 282
Sounds, percussion, 92
— special, **95**
— bell, **169**
— anvil, **159**
— succussion, 308
South Cornwall, 287
Spaces, intercostal, 6
Spasm, bronchial, 182
Specific fevers, 295
Special percussion sound, **95**
Spinal centres, 23
Spirometer, **116**

Spirometry, 116
Splenisation, 216
Spontaneous **pneumothorax, 165**
Sputum, colour, **56**
— composition, **56**
— consistence, **58**
— constituents, occasional, **58**
— how **to** examine, 58
— odour, **58**
— pneumonic, **222**
Sphygmograph, **119**
Sphygmography, **119**
Stages **of** whooping-cough, **192**
— of pneumonia, 215
Sternum, 3
Stethoscopes, 91
Stethometer, 114
Stethometry, 114
St Moritz, 287
Stimulants, cardiac, 44
— vascular, **44**
Structure of lung, 19
Subcrepitant râle, 102
Subcutaneous emphysema, 82
Subpleural tissue, 17
— tympanitic note, **17**
Succussion sound, 103, **169**
Sudamina, 81
Sudden death in paracentesis, 160
Suffocation, gradual, **50**
Suffocative breast pain, 67
Sulphate of copper, 284
Superior mediastinum, **10**
Symptoms, chest, 4
— influenza, 194
— hay asthma, **198**
— rose cold, 198
— emphysema, **205**
— mitral disease, **352**
— pulmonary disease, **358**
— congenital diseases of heart, 376
— aortitis, **381**
— pleurisy, **138**
Symptomatic treatment, 283, 373

27

Snycope, 75
— anginosa, 71
— in paracentesis, 159
Syphilitic phthisis, 290
— diagnosis, 292
— pathology, 290
— treatment, 293
— varieties, 290
Systole of heart, 34

T

Table of causes of dyspnœa, 52
Teneriffe, 287
Tension of pulse, 128
Terminations (*see* under each disease)
Thickened pleura, 284
Thoracic viscera, 45
Thrills, aneurysmal, 91, 389
— diastolic, 90
— pericardial, 91
— præsystolic, 90
— systolic, 90
Thoracic veins, diseases of, 400
Thrombosis, 279
Tinea versicolor, 81
Tone of percussion note, 92
Tonics, cardiac, 44
— vascular, 44
Too loud heart-sounds, 106
Too feeble heart-sounds, 106
Trachea, 11
Tracheal note, 92
Transfusion, 79
Transverse aortic arch, 45
— constriction of chest, 84
Traumatic pneumonia, 211
— pneumothorax, 266
Treatment (*see* under each head)
— of bronchiectasis, 189
— of catarrhal pneumonia, 235
— of cough, 60
— of croupous pneumonia, 227
— of emphysema, 209
— of hæmoptysis, 63

Treatment of influenza, 195
— of hay asthma, 198
— of palpitation, 66
— of hydrothorax, 162
— of pneumothorax, 192
— of whooping-cough, 192
— of rose cold, 198
— of interstitial pneumonia, 237
— of phthisis, 280
— of congenital disease of heart, 377
— aortitis, 381
Tricuspid valve, 30
— diseases of, 358
— — etiology, 358
— — signs, 358
Tri-nitrine, 74
Tripod of life, 47
Trocar and cannula for paracentesis, 156
Tubercular meningitis, 234, 279
Tubercle bacilli, 59, 270
— of heart, 339
— of pericardium, 308
— of pleura, 173
— true, 260
Tuberculosis, 279
Tubular breathing, 100
Tufnell's treatment, 391
Tumours, intrathoracic, 401

U

Ulceration of bronchi, 251
Ulcerative endocarditis, 327—333

V

Vagus, endings in lungs, 23
— action of, in respiration, 23
— — on heart, 37
— — in coughing, 54
— — in angina pectoris, 71

Valves of the heart, 29
— — mitral, 30
— — tricuspid, 30
— — semilunar, 31
— — coronary, 28
— — position of, 34
Valvular diseases, 342
— — aortic, 343
— — mitral, 351
— — pulmonary, 358
— — tricuspid, 358
Valvular murmurs, 107
Variations in heart-sounds, **106**
Varieties of pseudo-angina, **70**
Vascular sedatives, 44
— stimulants, 44
— tonics, 44
Veins of chest, 81
Veins, diseases of, **401**
Veine fluide, 107
Vena cava, 28
Ventnor, 287
Ventricle, right, 28
— left, 30
Villemin, **251**
Virchow, **267**
Viscera, thoracic, 45
Visceral pleura, 17
Vocal fremitus, 88
— — increased, 89
— — diminished, 89
Volume of pulse, 126
Vomiting in pneumonia, 220

W

Walls of chest, 1
— cavities, 263
Walshe, Dr., 362, 366
Wasting of chest-muscles, 82
— — — in phthisis, 275, 284
Water-hammer pulse, 350
Watson, **140**, 160
West, Dr S. H., 165, 300
West Indies, 286
Wetness of soil, 249
Whey, 285
Whooping-cough, **190**
— diagnosis, 191
— etiology, 190
— micro-organisms, 190
— **pathology,** 190
— **prognosis,** 192
— **signs,** 192
— stages, 190
— symptoms, 190
— treatment, 192
— terminations, 192
— **warning** of attack, **191**
Whooping-cough, 56
Width of intercostal spaces, 6
Wiesen, 287

X

Xiphoid cartilage, 3

Y

Yeo, Dr Burney, 252

University of Toronto Library

DO NOT REMOVE THE CARD FROM THIS POCKET

Acme Library Card Pocket
Under Pat "Ref. Index File"
Made by LIBRARY BUREAU

Author: Harris, Vincent D.
Title: Sinews of the Chest

H
M Phys
65015-

www.ingramcontent.com/pod-product-compliance
Lightning Source LLC
Chambersburg PA
CBHW020538300426
44111CB00008B/721